Palgrave Studies in Educational Futures

Series Editor
jan jagodzinski
Department of Secondary Education
University of Alberta
Edmonton, Alberta, Canada

"*Wild Pedagogies* leads an attack on the domestication of learning, the creeping formality, the structures and control, the standardizing, the separation of each of us from ourselves and our physical reality. Here's a portrait of the physicality of children and the need to embrace this in education; the need to see education as humans in the world; the need to experiment with ideas and words that give meaning to our intuitions. An important argument. An important book."
—John Ralston Saul, *award-winning essayist and novelist.*

"This seminal book is simply stunning. It offers educators and researchers genuine avenues—or what the authors call touchstones—for troubling dominant education systems. Jickling, Blenkinsop, Timmerman and Sitka-Sage offer "wild hope" in wild times where the planet has entered a new geologic era—the Anthropocene. Books like this are desperately needed to shift education into a new wild era where teachers are supported in returning to or opening up their own wildness in pedagogy. It is a return to what Leopold eloquently describes as "growing down," honouring wild childhoods."
—Amy Cutter-Mackenzie, *Sustainability, Environment & Education (SEE) Research Cluster, Southern Cross University, Australia.*

"Set in the most important of all contexts—with the future of life on our planet at stake—this book reminds us that whilst we can and must learn in and through wild (and perhaps not-so-wild) places, modern education systems around the world must also learn and adapt, and with urgency. I thoroughly recommend this book to anyone interested in education and its role in supporting the health of planet Earth—and surely that should mean all of us!"
—Peter Higgins, *Professor of Outdoor and Environmental Education and Director the Edinburgh Global Environment and Society Academy, University of Edinburgh, UK.*

The series Educational Futures would be a call on all aspects of education, not only specific subject specialist, but policy makers, religious education leaders, curriculum theorists, and those involved in shaping the educational imagination through its foundations and both psychoanalytical and psychological investments with youth to address this extraordinary precarity and anxiety that is continually rising as things do not get better but worsen. A global de-territorialization is taking place, and new voices and visions need to be seen and heard. The series would address the following questions and concerns. The three key signifiers of the book series title address this state of risk and emergency:

1. **The Anthropocene**: The 'human world,' the world-for-us is drifting toward a global situation where human extinction is not out of the question due to economic industrialization and overdevelopment, as well as the exponential growth of global population. How to we address this ecologically and educationally to still make a difference?
2. **Ecology**: What might be ways of re-thinking our relationships with the non-human forms of existence and in-human forms of artificial intelligence that have emerged? Are there possibilities to rework the ecological imagination educationally from its over-romanticized view of Nature, as many have argued: Nature and culture are no longer tenable separate signifiers. Can teachers and professors address the ideas that surround differentiated subjectivity where agency is no long attributed to the 'human' alone?
3. **Aesthetic Imaginaries**: What are the creative responses that can fabulate aesthetic imaginaries that are viable in specific contexts where the emergent ideas, which are able to gather heterogeneous elements together to present projects that address the two former descriptors: the Anthropocene and the every changing modulating ecologies. Can educators drawn on these aesthetic imaginaries to offer exploratory hope for what is a changing globe that is in constant crisis?

The series Educational Futures: Anthropocene, Ecology, and Aesthetic Imaginaries attempts to secure manuscripts that are aware of the precarity that reverberates throughout all life, and attempts to explore and experiment to develop an educational imagination which, at the very least, makes conscious what is a dire situation.

More information about this series at
http://www.palgrave.com/gp/series/15418

Bob Jickling • Sean Blenkinsop
Nora Timmerman • Michael
De Danann Sitka-Sage
Editors

Wild Pedagogies

Touchstones for Re-Negotiating Education and the Environment in the Anthropocene

palgrave
macmillan

Editors
Bob Jickling
Lakehead University
Thunder Bay, ON, Canada

Nora Timmerman
Sustainable Communities
Northern Arizona University
Flagstaff, AZ, USA

Sean Blenkinsop
Faculty of Education
Simon Fraser University
Vancouver, BC, Canada

Michael De Danann Sitka-Sage
Imaginative Education Research Group
Simon Fraser University
Burnaby, BC, Canada

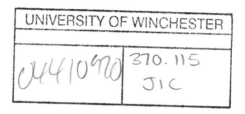

Palgrave Studies in Educational Futures
ISBN 978-3-319-90175-6   ISBN 978-3-319-90176-3 (eBook)
https://doi.org/10.1007/978-3-319-90176-3

Library of Congress Control Number: 2018943675

© The Editor(s) (if applicable) and The Author(s) 2018
This work is subject to copyright. All rights are solely and exclusively licensed by the Publisher, whether the whole or part of the material is concerned, specifically the rights of translation, reprinting, reuse of illustrations, recitation, broadcasting, reproduction on microfilms or in any other physical way, and transmission or information storage and retrieval, electronic adaptation, computer software, or by similar or dissimilar methodology now known or hereafter developed.
The use of general descriptive names, registered names, trademarks, service marks, etc. in this publication does not imply, even in the absence of a specific statement, that such names are exempt from the relevant protective laws and regulations and therefore free for general use.
The publisher, the authors and the editors are safe to assume that the advice and information in this book are believed to be true and accurate at the date of publication. Neither the publisher nor the authors or the editors give a warranty, express or implied, with respect to the material contained herein or for any errors or omissions that may have been made. The publisher remains neutral with regard to jurisdictional claims in published maps and institutional affiliations.

Cover illustration: Cover pattern © Melisa Hasan

Printed on acid-free paper

This Palgrave Macmillan imprint is published by the registered company Springer International Publishing AG part of Springer Nature.
The registered company address is: Gewerbestrasse 11, 6330 Cham, Switzerland

This was a unique project in that its core has been communally written while on a sailing boat off the West Coast of Scotland and during a writing retreat on the island of Luing, near Oban. It has been the task of the editors to gather this writing and weave its parts together into a single narrative. This book is dedicated to collaborative spirit and good will of its authors:

Hebrides, I., Ramsey Affifi, Sean Blenkinsop, Hans Gelter, Douglas Gilbert, Joyce Gilbert, Ruth Irwin, Aage Jensen, Bob Jickling, Polly Knowlton Cockett, Marcus Morse, Michael De Danann Sitka-Sage, Stephen Sterling, Nora Timmerman, and Andrea Welz.
Together these authors are referred to as the Crex Crex Collective—the taxonomical name for the migratory bird commonly called the Corncrake.

Attending to landscape and place has been a central part of this project. This book is also dedicated to the more-than-human contributors whose voices have made their way into this book through their own kind of authorship:

Bluebells, porpoises, the island of Luing, corncrakes, cuckoos, the Garvellachs, lava, quartz crystals, the Isle of Staffa, basalt columns, Fingal's cave, puffins, fulmars, shags, beachrocks, barren crags, seaside cliffs, the island of Iona, Lagandorain, Cnoc Buidhe (a small yellow hill), primrose, celandine, yellow flag iris, marsh marigold, birdsfoot trefoil, bog asphodel, sphagnum, grass, sea pinks, tussock grasses, silver sand eels, plankton, seaweed, crenulated cliffs, a creek, seals, good food, glacial erratics, moraines, beaches, Treshnish Isles, the sea, sea worn rocks, waves, wind, robins, Ardtornish bay, herring gulls, white-tailed eagles, sun, guillemots, razorbills, white sand, the island of Mull, otter, gneiss, schist, granite, currents, slate, sheep, dolphins.

Preface

In the 1970s my conceptions of schooling were upended. Living in Norway for a winter, I was astonished to learn that children there did not begin school until they were seven years old. If they attended any kind of pre-schooling they were expected to play—just play. Outside and in all kinds of weather. What a respite, I thought, for all those wiggling little bodies, to be spared the confinement of school. This wisely seemed to acknowledge the physicality of the children. Their foundational learning was sensuous, somatic, and experiential; they understood the world through being in it. As a young outdoor educator this held enormous appeal, mostly intuitive. And, like many others, I didn't have a very effective suite of words to talk about it.

Three decades later I returned to Norway for a "walking conference" that included a large number of local university students. Prompted by my earlier experience, I asked when they had started school. Interestingly, about half of them started at age seven, the other half at age six. They described this as a recent trend towards starting school earlier. During the same visit, I also arranged to visit a kindergarten located on a farm. As expected, the children were actively engaged outside. I was shown the chickens that they cared for; I watched them play in an adjacent forest; and, I even joined them for a short, and carefully monitored, goose-hunting excursion. Towards the end of the day, however, they returned to a small building near the barn where they had an hour of daily "academic" instruction. I was told that this academic preparation was a relatively recent addition; though, none of the teachers were entirely clear where the pressure for this emerging trend had arisen.

Pondering these experiences now, it seems important to find ways to talk more thoughtfully and deeply about the kinds of simple observations that struck me while in Norway. For myself and other keen observers this might begin by seeking words and ideas that describe our intuitions. To start, it seems very interesting that there are robust models of education that are physically active, play-infused, and located outside—in nature, or as we suggest in this book, in wild places. How important will this be for educational innovation? For education in general. But also, what are the consequences of the creeping formality, structure, and control of the learning as told in this story?

The Norwegian story isn't unique. Readers of this book will have their own experiences, intuitions, and stories. These will be stories about educational experiences bursting with potential. And sometimes they will, sadly, tell about how this potential is squandered as the experiences are bent to conform to more conventional expectations. Accordingly, many of you will have experimented with your own pedagogies—or maybe even in rearing your own children. Many of you have concerns about mainstream school systems; and, you see the educative potential of learning "outside" these systems. Indigenous people from around the world also have an abundance of concerns about how education has been controlled and imposed. And, they have their own stories about how education had been traditionally tied to the land, and about how these connections have been severed. Collectively, stories will intersect and overlap. As Robert Bringhurst has so beautifully written, stories tend to "sprawl all over each other."[1] This book is an invitation to celebrate a multiplicity of stories—to experiment with them, and to see what work they can do for you, and even write some new ones. This book is also an invitation to experiment with ideas and words that reflect points of collective convergence and give meaning to our intuitions. Here, we nominate the term "wild pedagogies" as a lens for these converging stories, and as a guide to enable our experiments in practice.

What we propose is no idle debate. As David Orr has maintained, "Our science and technology have changed beyond recognition and the Earth is rapidly shifting from the Holocene to something being called the

Anthropocene. Our descendants, assuming they exist, will live on what Bill McKibben calls 'Eaarth,' a more capricious and threadbare planet."[2] Education, of course, cannot by itself resolve the dire warnings inherent in Orr's observation. However it does have a vital role to play. We agree with Orr when he goes on to say, "Without exaggeration it will come down to whether students come through their formal schooling as more clever vandals of the Earth and of each other or as loving, caring, compassionate, and competent healers, restorers, builders, and midwives to a decent, durable, and beautiful future."[3] Framed this way, it is clear to us that serious educational change is required.

What then will it take to nurture healers and restorers of the earth rather than evermore clever vandals? "We will not solve problems using the same thinking that created the problems in the first place" is an old meme that, as is often the fate of popularized slogans, feels cliché. However, Audre Lorde added substance to this notion when she said, "*the master's tools will never dismantle the master's house. They may allow us temporarily to beat him at his own game, but they will never enable us to bring about genuine change.*"[4] Perhaps we could rework these thoughts a little further, to better align with the concerns of this book. What if we switch them up to become: "We won't nurture compassionate healers and restorers, if we continue to *be* the same people that have vandalized our Earth."

Years ago, Joe Sheridan wrote evocatively about the significance of walking as pedagogy.[5] Walking in a landscape, using mind and feet in concert, he reminded us, is the oldest educational method. Walking is the most fundamental pedagogy known to humans. Humans are physical, sensuous beings; and, we learn through this physicality—though our lived experiences in the world. Through *being in the world* we learn differently, and we learn to be different people. For those who have walked throughout the landscape of rationalist traditions much of our lives, we know it is one thing to glean inklings about the world by looking through the windows provided by mainstream education. It is quite another thing to stand on the fertile earth of learning. Sheridan reminds us, by looking back at ancient pedagogical traditions, how much mainstream education has become constrained and controlled. We hope that now, in going forward, wild pedagogies will help us all to find a better footing—literally and metaphorically.

Sheridan also warns us of the dangers of *domesticating* education. For him, domestication means managing the life right out of it—taming it, restraining it, confining it, controlling it. Education, as is it most often encountered—that is, inside, seated, standardized, and more-or-less still—is a world of abstraction and heavily, perhaps even oppressively, mediated experiences. Real life is outside—literally and educationally. Compliance with these norms of indoor study denies our physicality. Sheridan likens this to breaking horses, the violent expunging of our inherent wildness.

This isn't just rhetoric. The Norwegian example shows how educational practices can be slowly broken down, and bent towards evermore-tamed versions of themselves. More broadly, research[6] illustrates a broad trend towards prescribed curricula aimed at teaching to standardized tests. But, it is not just curricular content that is subject to standardization. The lived practices of teachers, the idea of who teachers are, and the places where those practices can occur are repressed by the weight of expectations—wildness is unwelcome. This works against allowing education to dwell in the world. To *be* different.

Wild pedagogy is aimed at "re-wilding" education. In this aim is an acknowledgement that learning to become different people will require being in the world—dwelling there. Re-wilding education thus requires learning from place and landscape. Listening to voices from the more-than-human world. Attending to the untamed. This will require, at the least, making the walls around modern western education more permeable and in some cases, removing them all together. The alternative is to double down on the walls, literally and figuratively. We have seen this during the past couple of decades and it can take the forms of: "academic preparation" being extended to six, five, four, and even three-year olds; zealous demands for accountability; narrowly restrictive standards of practice; and, fanciful notions of the "basics." Typically these prescriptions are enacted in the name of rigour.

Rigour, however, is derived from stiffness, rigidity, and is a short step from death, as in *rigour mortis*. Let's not forget that education can become part of the walls that we build around our lives. And, when those walls become rigid they become the tombs we build around ourselves.[7]

Fortunately, all is not quite so lifeless as it may seem. Research also highlights teachers who, in spite of testing pressure, insist on creating space for what they consider "real teaching." There is resurgence in nature-base education, especially following publication of Richard Louv's book, *Last Child in the Woods*.[8] The Forest School Movement for young children is growing and spreading beyond its European origins. Alternative and integrated programs thrive in many school districts. There is even one Canadian school in Maple Ridge that has no permanent school building, preferring to conduct most instruction outside in relatively "wild places."

The preceding text outlines some of the personal seeds that have grown into an idea we call wild pedagogies. In the last few years, this idea has generated considerable interest through small conferences, gatherings, and at international venues. *Wild Pedagogies* began as a graduate course at Lakehead University in 2012. Then, in the summer of 2014, a broad cross-section of educators gathered on the Yukon River for a first conference, *Wild Pedagogies: A Floating Colloquium*. The narratives generated by this initial run were published in a special issue of the outdoor education journal *Pathways*.[9] The conversation has been carried along through Conference presentations such as at World Environmental Education Congresses, in 2015 and 2017, and the annual meeting of the Canadian Network for Environmental Education and Communication in 2016. Additionally, there have been some ad-hoc gatherings, including: *Wild Pedagogies: The Tetrahedron Dialogues*, in May 2016. The name seems to resonate with many and its persistence suggests that there must be more to it than just a catchy slogan. Its long-term viability and usefulness demands the same.

As successful as the earlier wild pedagogy gatherings have been, a number of participants realised that there was more work to do and this work needed to be responsive to two themes. First, tracings of the wild pedagogy idea have been scattered throughout small publications and group adventures. Collectively, however, a coherent vision of wild pedagogies, with some theoretical and practical depth, remained slim. This book was proposed to add depth to the idea—to take it beyond a catchy slogan and give it life. But, how could this be done?

The first clue lies in the pluralisation of wild pedagogies. While this book aims to locate emergent themes and give the ideas some coherence, we are convinced that there will never be—nor should there be—a single wild pedagogy. Second, the richness and diversity within an idea can be greater when more people have a stake in its development and emerging practices. To put it another way, this was a perfect opportunity to experiment with eco-social learning and to make meaning in a collaborative setting.

With these clues in mind, this book is the result of an intentionally collaborative project. A group of international adventurers, teachers, and scholars, and the more-than-human beings they encountered, met together during a *Wild Pedagogies Sailing Colloquium* during May 2017. This gathering took place off the West Coast of Scotland on a large sailboat, *The Lady of Avenel*, and on the small island of Luing near Oban. This book has arisen out an experiment with interested folks, educational ideas, and wild places. By *being in the world* we aspired to do our work differently—to think, learn, and write differently—and, perhaps, to learn to be different pedagogues.

This book also aims to lend a supporting voice to educators who are already wild pedagogues. It aims to find threads that run through the sprawling and intersecting stories that these educators tell. It suggests touchstones that educators can return to when they wish to refresh their teaching, or when they feel pressure to bend towards normalized conventions. It also aims to nurture conversations amongst those who recognize that education, writ large, is failing the planet, the resident beings, and themselves. It aims to provoke conversations in public places as people everywhere grapple with failing educational systems, competing ideologies, and the co-option of schooling by neoliberal interests, amongst others.

As Jay Griffiths reminds us, "All things that represent life at its most vital and wild wiggle. Words wiggle into metaphor; sperm wiggles; dancing and jokes and giggling wiggle; the shape and character of tumultuous life is a wiggling one."[10] In the chapters that follow, we too shall wiggle our way through conceptual, poetic, and philosophic experiments in an attempt to articulate wild pedagogies. In doing so, we invite you, in your own ways, to wiggle with us and give shape and character to the tumultuous lives we share on this precious planet.

Thunder Bay, ON, Canada   Bob Jickling
January 2018

Notes

1. Robert Bringhurst, "The Tree of Meaning and the Work of Ecological Linguistics," *Canadian Journal of Environmental Education* 7, no. 2 (2002): 9–22.
2. David Orr, "Foreword," in *Post-Sustainability and Environmental Education: Remaking Education for the Future*, ed. B. Jickling and S. Sterling (London: Palgrave Macmillan, 2017): ix.
3. Ibid., ix–x.
4. Audre Lorde, "The Master's Tools Will Never Dismantle the Master's House," in *This Bridge Called My Back: Writings by Radical Women of Color*, ed. Chertrie Moraga and Gloria Anzaldúa (New York: Kitchen Table Press, 1983): 94–101.
5. Joe Sheridan, "My Name is Walker: An Environmental Resistance Exodus," *Canadian Journal of Environmental Education* 7, no. 2 (2002): 193–206.
6. Research will be examined more thoroughly within the text of the book. Also, much research alluded to here is captured in: *Post-Sustainability and Environmental Education: Remaking Education for the Future*, eds. B. Jickling and S. Sterling (London: Palgrave Macmillan, 2017). Also: Smith, W.C. (Ed.) (2016). *The Global Testing Culture: Shaping Educational Policy, Perceptions, and Practice*. Oxford: Symposium Books.
7. This idea riffs off of George Elliott Clarke's *Whylah Falls* (Polestar Press: Winlaw, British Columbia, 1990): 49. And this is a reminder of the importance of poetry and fiction as research.
8. Richard Louv, *Last Child in the Woods* (Chapel Hill: Algonquin Books, 2005).
9. *Pathways: The Ontario Journal of Outdoor Education* 28, no. 4 (2016).
10. Jay Griffiths, *Wild: An Elemental Journey* (New York: Jeremy P. Tarcher, 2006): 66.

References

Bringhurst, R. "The Tree of Meaning and the Work of Ecological Linguistics." *Canadian Journal of Environmental Education* 7, no. 2 (2002): 9–22.
Clarke G.E. *Whylah Falls.* Winlay: Polestar Press, 1990.
Griffiths, J. *Wild: An Elemental Journey*. New York: Jeremy P. Tarcher, 2006.

Jickling B., and S. Sterling, eds. *Post-Sustainability and Environmental Education: Remaking Education for the Future*. London: Palgrave Macmillan, 2017.

Lorde, A. "The Master's Tools Will Never Dismantle the Master's House." In *This Bridge Called My Back: Writings by Radical Women of Color*, ed. C. Moraga and G. Anzaldúa, 94–101. New York: Kitchen Table Press, 1983.

Louv, R. *Last Child in the Woods*. Chapel Hill: Algonquin Books, 2005.

Orr, D. "Foreword." In *Post-Sustainability and Environmental Education: Remaking Education for the Future*, ed. R. Jickling and S. Sterling, vii–xi. London: Palgrave Macmillan, 2017.

Sheridan, J. "My Name Is Walker: An Environmental Resistance Exodus." *Canadian Journal of Environmental Education* 7, no. 2 (2002): 193–206.

Smith, W.C., ed. *The Global Testing Culture: Shaping Educational Policy, Perceptions, and Practice*. Oxford: Symposium Books, 2016.

Acknowledgements

This book willed itself into existence through gathering a briny pod of international scholars on a sailing boat off the West Coast of Scotland for a *Sailing Colloquium*. The writing caught more wind in its sails during a contiguous retreat on one of the small Slate Islands called Luing. Creating such a context for writing a book was no small feat. However we did have some help and would like to acknowledge and thank Joyce and Douglas Gilbert for their support, without which this would never have happened.

Joyce's boundless enthusiasm, cultural knowledge, and experience running trips like ours were invaluable. She located our ship, the *Lady of Avenel*, retreat location, the Atlantic Islands Centre, and special guest speakers and performers. Then, throughout the colloquium, she was a full participant in discussions, often infusing conversations with insightful local knowledge.

Douglas was our resident ecologist and natural historian and, as such, he was ever present to interpret the Scottish landscape and its inhabitants. But there was much more. Douglas's obvious love of the places through which we travelled often led to insightful vignettes as he drew our attention to wondrous little events that were taking place, often unnoticed by the rest of us. Like Joyce, Douglas engaged fully in other colloquium activities and discussions.

We also acknowledge, and thank, Wendy Jickling. She took on the challenging task of feeding us. Not only did this involve planning menus and ordering food, but also preparing meals in the small galley of a sometimes-moving ship. The results were always superb. As a lover of Scotland and a retired school-teacher, herself, she also brought her inner wild pedagogue to the group, every day.

The *Lady of Avenal* was our home, companion, and more-than-human interlocutor. Christopher J. Wren was her Skipper and Tom Lagan her mate. We thank Chris and Tom for their wisdom and skill, but also their companionship throughout and their ready willingness to embrace the spirit of wild pedagogies. We also acknowledge Stefan Fritz and Jane Emery for their support throughout and ensuring that the *Lady* was ready and waiting for us in Oban.

What a wonderful experience to live on Luing, for a time. It was pleasure to meet Zoe and Alistair Fleming, Norrie Bissell, Birgit Whitmore, and Alison Robertson on their island. And, we thank them for continually going out of their way to ensure that our every need was met.

While on Luing our Scottish experience was deepened through artistic performance, and a rousing ceilidh. Islander Norman Bissell shared poems about his journey to Luing, he also talked with us about his vision for geo-poetics. Claire Hewitt is an extraordinary harpist, singer, and storyteller; her performances were thrilling links to the places of our visit. Bob Pegg's stories and songs dug deeply into Neolithic pre-Scotland, particularly through his enthusiasm for playing simple little whistles made from tiny stones and bones.

We also thank Robbie Nicol for making the journey from the University of Edinburgh to Luing in order to meet with us, share his expertise, and talk about wild pedagogies.

While on Luing, the Atlantic Islands Centre was a hub of our activity. We were entertained there, we sometimes ate there, we often worked there, and at times it felt a bit like we overran the place. We are indebted to the staff at the centre for their excellent food, willingness to host us, and the kind support that they offered throughout our time there. Thank you: Archie Frost, Rebecca Robertson, Rikki-Lee Burton, Kirsty Gommersall, Eugenie Thomas, Rachel Cruickshanks, Rhona Mackenzie, Ashleigh Ord, Imogen Rennie, and Erin Walshaw.

Walking to the store on Luing was sometimes answering a calling for an outing, a break from our deliberations, or the pleasure of the journey. And, we even got supplies there. Gordon Robbie and Simone Van Dijl, like everyone on Luing went out of their way to ensure that we had everything we needed.

From another island, Iona, John MacLean kindly helped to clarify some little points in the story about the "yellow hill." And in Canada, Paddy Blenkinsop kindly cast a critical and insightful eye across Chap. 5.

Finally, we have included in the final chapter of this book three stories from participants from the earlier: *Wild Pedagogies: A Floating Colloquium*. We acknowledge and thank Victor Elderton, Yuko Oguri, and Vivian Wood-Alexander for their enthusiasm in embracing wild pedagogies for bravely sharing their stories. We also gratefully acknowledge permission to reproduce the poem "These Sea-Worn Rocks" from *Slate, Sea and Sky* by Norman Bissell, Edingurgh, Luath Press.

Contents

1 Why Wild Pedagogies?   1

2 On Wilderness   23

3 On the Anthropocene   51

4 On Education   63

5 Six Touchstones for Wild Pedagogies in Practice   77

6 Afterwords   109

Index   133

The Crex Crex Collective

The Crex Crex Collective includes: Hebrides, I., Independent Scholar; Ramsey Affifi, University of Edinburgh; Sean Blenkinsop, Simon Fraser University; Hans Gelter, Guide Natura & Luleå, University of Technology; Douglas Gilbert, Trees for Life; Joyce Gilbert, Trees for Life; Ruth Irwin, Independent Scholar; Aage Jensen, Nord University; Bob Jickling, Lakehead University; Polly Knowlton Cockett, University of Calgary; Marcus Morse, La Trobe University; Michael De Danann Sitka-Sage, Simon Fraser University; Stephen Sterling, University of Plymouth; Nora Timmerman, Northern Arizona University; and Andrea Welz, Sault College.

Crex crex is the taxonomical name given to the Corncrake. We have chosen this bird to represent our collective because it was an important collaborator in this project and because its onomatopoeic name beautifully mirrors its call—a raspy crex crex. For some reason, it chooses to fly over England and breeds in Scotland and Ireland. Presumably this is due to loss of habitat in modern England, but perhaps these birds sense some epicenter of empire there? Who is to know?

Notes on Contributors

Ramsey Affifi is a Teaching Fellow in Outdoor, Environmental, and Sustainability Education at the University of Edinburgh. He is founder of the organizations Sustainable Laos Education Initiatives, and The Sai Nyai Eco-School. Ramsey's passions are learning and teaching, and being with plants and birds. He is currently exploring the emotional/existential dimensions of the ecological crisis, and the pedagogical value of humility, awe, and grief.

Sean Blenkinsop is Professor, Faculty of Education, Simon Fraser University, Vancouver, Canada. He grew up in the boreal forests of northern Canada. With a 30+ year background in outdoor, environmental, and experiential education his interest in wild pedagogies comes quite naturally. Now, as Professor of Education he has been involved in starting three nature-based, place-based, eco-schools (all in the public system) and has written extensively about these experiences and the philosophical underpinnings of eco-education writ large.

Michael De Danann Sitka-Sage's interests in environmental education… springs from a background in eco-anarchic politics, but also a profound love of the humanities and a fortunate childhood spent in the backcountry of "Vancouver Island" on the traditional territories of the K'omoks. For the past six years, he has been a member of the SFU research team involved with the Maple Ridge Environmental School. His interest in the Wild Pedagogy movement lies in pushing the boundaries of what conferences can be and what provokes thinking with swashbuckling and wild folk.

Hans Gelter having a PhD in biology, has been teaching biology, environmental education and outdoor education for over 32 years. In different educational programs and courses, he has developed what can be called "wild pedagogics," with field trips in Northern Scandinavia, Nepal, and Antarctica. In is his ecotourist company Guide Natura, Hans does guiding and interpretation of the coastal environment of Swedish Lapland, has a kayak centre, and drives a tour- and taxi boat.

Douglas Gilbert grew up in the not-so-wild streets of Edinburgh, dodging taxis on his bicycle and dreaming of being in the Scottish mountains. It took him many years to crystalise an enthusiasm for climbing hills into a love of nature, via a degree in Zoology. He currently works for Scottish conservation charity Trees for Life, promoting native forest restoration and re-wilding Dundreggan Conservation Estate near Loch Ness (https://treesforlife.org.uk/work/dundreggan). He has previous experience for the Royal Society for the Protection of Birds as an ecologist, and has worked for government agencies promoting nature conservation in Scotland and England over a span of 30 years in various roles. A keen naturalist, he has acted as a wildlife guide on many trips on the west coast of Scotland.

Joyce Gilbert originally earned a PhD in biochemistry, but gave all that up to be an environmental educator. During the last couple of decades, she has acquired enormous experience across a broad range of educational settings, including formal schooling and with Non-Governmental Organisations. Joyce has run many on-board courses and workshops off the West Coast of Scotland. And, with her knowledge of local cultural history, and with her studies in the Gaelic language, she was a key collaborator on this project. Joyce currently works for the Scottish Non-Governmental organization Trees for Life. It is a small conservation charity that seeks to restore the ancient Caledonian forest in the Highlands of Scotland.

I. Hebrides is an independent scholar with deep roots in the culture and ecology of Western Scotland. She has expertise relating to historical and current ways of living and being in this sparse island filled ocean landscape. Her insights have had an important influence on many resident communities in the area and she frequently collaborates with researchers within Scotland and abroad. She was excited and indifferent, often at the same time, to engage with all the other invitees of the sailing colloquium. And sometimes she speaks on behalf of the more-than-human and material worlds on the west coast of Scotland.

Ruth Irwin is a Kiwi by birth… and a rather nomadic person, now living in Scotland again. From one set of isles to another, the first set on tectonic upsweep, young steep sharp jaggedy mountains, and deep dark forest, and always a heartbeat from sand and sea. In Scotland the mountains are of a different order altogether. Its old here, prehistoric, with worn granite curves that have weathered down over such an age, that each mountain is like an inverted bowl, steep at the base and shallow curves, voluptuousness at the crest. The forest is multi-coloured at this time of year; gold, and brown and silvery greens. With sea lochs diverting inland, and the route from coast to coast as narrow as New Zealand.

Her research is on climate change, philosophy, and education. She has authored books such as *Heidegger, Politics and Climate Change* (2008). Right now she is working on a text looking at climate change, economic de-growth, and the debt jubilee from Mesopotamia. She tends to view modernity from an anthropological perspective, and is very interested in philosophy of technology, place, and time.

Aage Jensen grew up in the Norwegian countryside in the late nineteen fifties and had a close relationship with the forests, mountains and the sea. He is educated as a teacher and has practised in all levels in the Norwegian school system. In the last fifteen years before retirement was an Assistant Professor at Nord University in Norway. He has degrees in both biology and pedagogy.

Bob Jickling worked as a Professor of Education at Lakehead University after many years of teaching at Yukon College. He continues his work as Professor Emeritus. He now lives in Canada's Yukon Territory because, from the beginning in 1979, this land has sung to him. And it sang while writing this statement when a museum of Bohemian Waxwings cruised through the neighbourhood. Eventually they descended on all of the Mountain Ash trees stripping them of the berries. Living, in such a place, he has never abandoned the idea of wilderness. It cannot simply be written off as a social construct—the waxwings are real and they have agency. He has puzzled about this for years, but more recently has found resonant chords in his concerns about education. Along the way, he founded *Canadian Journal of Environmental Education*, co-chaired the 5th World Environmental Education Congress in Montreal, and wrote many scholarly papers. But now he has returned to his Yukon home, to hear it sing, to complete a few projects, and to ponder possibilities for "wild pedagogies."

Polly Knowlton Cockett with a background in geology and environmental education, is involved in ecological literacy, sense of place studies, urban native biodiversity conservation, and socioecological cartography in the high prairies of Alberta. A former teacher, she currently works as an Instructor in the Werklund School of Education at the University of Calgary, serves on the BiodiverCity Advisory Committee for The City of Calgary, and as President of Grassroutes Ethnoecological Association is engaged in creative community place-making endeavours.

Marcus Morse grew up in Tasmania where he spent a great deal of time journeying in its many wild places. He is currently a senior lecturer in outdoor environmental education at La Trobe University in Victoria, Australia. Marcus' teaching and research focus on developing imaginative educational approaches that invite students to engage with places and people in ways that pay attention to the wildness, vibrancy, and complexity of the world.

Stephen Sterling has been working for the last 40 years or so on trying to articulate, demonstrate, and advocate an approach to education that is deeply relational and sustains people and planet in a harmonious state. The colloquium reflected this challenge, and manifested its pursuit of wild pedagogy by pushing our thinking, and by stretching our experience of nature and landscape—going beyond the norm to realise a living educational practice.

He is Emeritus Professor at the University of Plymouth and a former Senior Advisor to the UK Higher Education Academy on Education for Sustainable Development (ESD) and National Teaching Fellow (NTF), and have worked in environmental and sustainability education in the academic and NGO fields nationally and internationally for many years. He is widely published: the first book (co-edited with John Huckle) was *Education for Sustainability* (Earthscan, 1996), and this was followed by the influential Schumacher Briefing *Sustainable Education—Re-visioning Learning and Change* (Green Books 2001). The most recent book, with Bob Jickling, is *Post-Sustainability and Environmental Education: Remaking Education for the Future*, Pivot Press/Palgrave (2017). His research interests centre on the interrelationships between ecological thinking, systemic change, and learning at individual and institutional scales to help meet the challenge of accelerating the educational response to the sustainability agenda.

Nora Timmerman grew to adulthood in the dry heat of Arizona, subsequently spent nine years in the wet, cool of Vancouver, Canada, and five years ago moved back to Arizona—but to the high, mountainous regions. Living now within the largest contiguous Ponderosa pine forest, Nora revels in exploring the wildness of life on the Colorado Plateau. She also finds much wildness and wonder in her family, community, garden, dancing, and work. Nora is a Lecturer at Northern Arizona University in the Sustainable Communities Program. In this program, her teaching brings students, community partners, and faculty together to learn and organize toward just and sustainable communities. Nora's research and writing focus particularly on how postsecondary education faculty practice and collectively strive for socio-ecological justice with/in postsecondary education contexts.

Andrea Welz as an educator and parent, feels that connections with the rest of the natural world are an important part of human development and a way to foster ecological consciousness. Her memories of a childhood with the freedom to roam the fields, streams, ponds, and forests near her home have led to her to investigate nature-based early childhood programs in Europe and North America. Wild pedagogy offers a fresh new perspective that is broadening her understanding of connections with nature in early childhood, and how these connections can be nurtured. She is integrating these thoughts in her work as an Early Childhood Educator Program faculty member at Sault College, Canada, and in the development of a nature-based preschool program with Little Lions Child and Family Centre.

List of Images

Image 1.1	Wind change in the rigging. Photo credit: Aage Jensen	10
Image 1.2	The *Lady of Avenel* lying off on Iona. Photo credit: Bob Jickling	14
Image 1.3	Ancient stone cell on *Eilach an Naoimh*. Photo credit: Hansi Gelter	17
Image 2.1	*Cnoc Buidhe* (Yellow Hill). Photo credit: Bob Jickling	28
Image 2.2	A Puffin on the Island of Staffa. Photo credit: Hansi Gelter	42
Image 3.1	Sands of time. Photo credit: Hansi Gelter	54
Image 4.1	A seal presence. Photo credit: Hansi Gelter	73
Image 5.1	Bluebells blooming. Photo credit: Hansi Gelter	81
Image 5.2	Basalt columns on Staffa. Photo credit: Hansi Gelter	85
Image 5.3	The Lady of Avenel in the Garvellachs. Photo credit: Hansi Gelter	93
Image 5.4	Voices amongst us. Photo credit: Hansi Gelter	96

List of Boxed Vignettes

From the Ship's Manifest: Wild Pedagogies Version	9
Eilach an Naoimh, in the Garvellachs	16
Cnoc Buidhe (Yellow Hill)	27
Puffins	41
Fulmar Petrels	57
Stepping Out	68
A Welcome Interruption	72
THESE SEA-WORN ROCKS	106

1

Why Wild Pedagogies?

The Crex Crex Collective

Abstract Given the sense of ecological urgency that increasingly defines our times, this chapter seeks to look beyond current norms and worldviews that are environmentally problematic. With this thinking in mind, wild pedagogies, first, aims to re-examine relationships with places, landscapes, nature, more-than-human beings, and the wild. This requires rethinking the concepts wilderness, wildness, and freedom. Second, this chapter contends that educators need to trouble the dominant versions of education that are enacted in powerful ways and that bend outcomes towards a human-centred and unecological *status quo*. With this in

---

The Crex Crex Collective includes: Hebrides, I., Independent Scholar; Ramsey Affifi, University of Edinburgh; Sean Blenkinsop, Simon Fraser University; Hans Gelter, Guide Natura & Luleå, University of Technology; Douglas Gilbert, Trees for Life; Joyce Gilbert, Trees for Life; Ruth Irwin, Independent Scholar; Aage Jensen, Nord University; Bob Jickling, Lakehead University; Polly Knowlton Cockett, University of Calgary; Marcus Morse, La Trobe University; Michael De Danann Sitka-Sage, Simon Fraser University; Stephen Sterling, University of Plymouth; Nora Timmerman, Northern Arizona University; and Andrea Welz, Sault College.

Bob Jickling (bob.jickling@lakeheadu.ca) is the corresponding author.

B. Jickling (✉)
Lakehead University, Thunder Bay, ON, Canada
e-mail: bob.jickling@lakeheadu.ca

mind, wild pedagogies seeks to challenge recent trends towards increased control over pedagogy and education, and how this control is constraining and domesticating educators, teachers, and the curriculum. Finally, given that the dominant current human relationship with Earth cannot be sustained, we posit that any critique suggested must be paired with a vision—and corresponding educational tools—that allows for the possibility of enacting a new relationship.

Keywords Anthropocene • Control • Education • Environment • More-than-human • Pedagogy • Wild • Wilderness

The earliest experiments with wild pedagogies were, at their core, about reimagining and enacting alternative relationships. By alternative, we mean relationships that fall outside of mainstream business, politics, and education. Given the sense of ecological urgency that increasingly defines our times, it seems important to look beyond present norms and worldviews for our responses. With this thinking in mind, our first aim was to re-examine relationships with places, landscapes, nature, more-than-human beings, and the wild. This required rethinking notions of wilderness, wildness, and freedom. The second aim was to challenge recent trends towards increased control over pedagogy and education, and how this control has been constraining and domesticating. The third aim was to offer something to educators—something that would propose a possible path forward. This book thus builds on past work with the aim of more thoroughly articulating the theoretical roots and offering practical strategies for enacting wild pedagogies.

It is tempting to say that Earth is in terminal decline. Climate change appears to have reached, or perhaps even crossed, the threshold of irreversibility. Many scientists and environmental historians are carefully arguing that Earth is about to, or is already, transitioning into a new geological epoch: the Anthropocene, or the "age of human impact." If we are indeed at the threshold of a new epoch, we are equally at the threshold of an emerging geological conversation. A new geo-story is transpiring on the ground beneath our feet, in the atmosphere around us, and in increasingly warm oceans. It is being communicated through climate change, increasingly fierce storms, and mass species extinctions. The first to suffer, as always, are the marginalized and disenfranchised. This includes many from our human species—especially within oppressed communities—but also the staggering loss and suffering of other species. This suffering of more-than-human beings seldom registers in public discourse and is often downplayed. However, the situation appears to be worsening. Science is inherently sceptical, cautious, you could even say conservative. Yet, with each passing year we hear news reporting that the situation is even worse than previously predicted—more species annihilated, more glacial ice melted, and more topsoil lost to the sea.

As frightening as this is, Earth has seen large-scale extinction events before. In fact, there appears to have been five of them. If another drastic change is imminent—in geological time—Earth will survive. It has before. Many existing species will not; others will be diminished—and this includes humans. Yet, it is not humans as a whole that are the source of this problem. So what is the trouble? We argue, what is really at issue is a troubling kind of relationship with Earth. This relationship is reflected in the ways that many of the most affluent and "developed" nations have lost the knowledge of and, subverted the social structures for, living well with place—to live within their means, to live with care and compassion for other beings, and to live with wonder in Earth itself. Unfortunately, this relationship appears to be spreading across the globe.

We wonder what the world could look like if humans, afflicted with such relationships within their place on Earth, enacted different ways of being in the world. What is not clear is how, exactly, it would mean to "be differently" within the world. It seems that the most common suggestions, which range across various new attitudes, prescriptions, warnings, restrictions, summons, sermons, and threats, seem to be out of sync with the magnitude of changes required. Changing relationships within our place on Earth or being in the world differently entails far more than using a different kind of light bulb. Something more fundamental must be disrupted[1] and something significantly different must be offered. Such a disruption will not be achieved through appeals to rationality, duties, or facts alone, nor will it be achieved by humans on their own. It is more likely that changing relationships with Earth and its other beings will require learning through active engagement with the natural world. The return could be rich—for example, in increased sense of well-being, decreased sense of alienation, and an expanded range of what it means to be human. Thus, this conversation about change is about doing, not just thinking. And, doing things differently will mean being different in our orientations towards nature, our language about nature, and our responsibilities with nature. This suggests that we must practice a kind of environmental etiquette. Here etiquette is not reserved for elites, but it is rather a kind of everyday manners amongst beings and places. This move to being differently in world also suggests that we are tasked to engage in face-to-face "re-negotiating" of what it means to be human and to be a citizen in a more-than-human world. All of this is, at least in part, an educational project. Big changes are needed and with big changes bold educational approaches are required. In this book we propose some bold—and wild—ideas.

Before launching into wild pedagogies, we need to acknowledge the way in which education has typically been conceived is in trouble. Kris Gutiérrez,[2] former President of the American Educational Research Association, provided a good summary of this problem when she described her most persistent concern:

> Our inability to intervene productively, at least in any sustained and transformative way, in the academic lives of so many youth today—to imagine new trajectories and future forms of agency…. we simply cannot rely on efficiency and market-driven models of education that are certain to bankrupt the future of our nation's youth. We need models for educational intervention that are consequential—new systems that demand radical shifts in our views of learning….

For Gutiérrez, current models of education are part of the problem. Her vision of education demands that she enact pedagogies that challenge dominant models. In doing so, she is simultaneously being the change in the very process of enabling change to occur.

Gutiérrez's own pedagogy reflects a view that learning and developing agency requires doing, engaging, being the change—and, indeed, being in the world differently. Her own educational experiments point to wonderful pedagogical possibilities when university students and school children work together. In her words, "They were brought together through an intervention that privileged joint activity, playful imagination, and a vision of teaching in which an imagined or projected future could influence activity in the present."[3]

We prefer to think of Gutiérrez's pedagogies as well-crafted experiential learning opportunities—rather than interventions. Here we draw attention to the word "intervene." Intervention implies that we can actually pinpoint a problem and, through a diagnostic process, treat the condition and stipulate the outcome. Our worry is that the language of "interventions" can easily slip into a good idea gone awry whereby, in this case, humans educators are seen as "thinkers and fixers"—they remain in control. This does not appear to be what Gutiérrez does in the wonderful examples she offers from her own practice. However, the persistence of this kind of language and thinking, in more general educational conversations does not, we fear, offer the kinds of resistance that can enable entry into a bold alternative discourse.

Consider UNESCO's response to a need for educational change. Irina Bokova,[4] the (former) Director-General of UNESCO sheds light on the persistence of the educational problem, as we see it. In the foreword to her agency's recent report, *Education for People and Planet: Creating Sustainable Futures for All*, she asserts, "we must fundamentally change the way we think about education and its role in human well-being and global development." However, she offers the same tired old rhetoric. For her, educators have a responsibility "to foster the right type of skills, attitudes and behavior that will lead to sustainable and inclusive growth."[5] This suggests, to us, that Bokova is proposing an interventionist strategy that is, in all likelihood, not adequate for achieving the deeper and, indeed, transformational change that we believe is required. Sociologist, Zygmunt Bauman challenges this conventional, and indeed delusional, vision of educational "innovation."

Bauman[6] claims that approaches like Bokova's are habitual and tired answers to "the wrong kind of behavior." Bauman is doubtful that social realities can be unsettled, dislodged, or even radically changed by simply attempting to instil in learners "new kinds of motives, developing different propensities and training them in deploying new skills."[7] He insightfully asks whether educators performing in this way will ultimately "be able to avoid being enlisted in the service of the self-same pressures they are meant to defy?"[8] This question goes to the heart of the task before us. And, a further look at Bokova's comment, as a common example, underscores key issues.

In the first place, nothing in Bokova's statement suggests a need to fundamentally disrupt, and re-negotiate humankind's relationship with Earth. As Stephen Sterling observes, "UNESCO has been suggesting the need for a 'new vision' of education for some time yet what is often missing is a sufficient critique of the dominant cultural worldview."[9] Without such a critique it is hard to imagine how UNESCO's initiatives will do anything other than succumb to the *status quo*, as Bauman warned. Second, Bokova's interventionist solution assumes humans can somehow control and correct their own fate without accounting for the volatility and turbulence of our times, or the role the natural world is likely going to have to play. Absent from the United Nations and UNESCO's discourse is any serious reflection about what education is, or could be.[10] Implicit in this omission is an assumption that mainstream education as presently conceived is, for all intents and purposes, largely adequate. Finally, Bokova's statement is all about humans—particularly human well-being that is linked to global development. There is no concern for what well-being means in the more-than-human world and no suggestion that Earth can have educational agency.

We contend that educators need to trouble the dominant versions of education that are enacted in powerful ways and that bend outcomes towards the *status quo*. But, troubling dominant language, norms, and practices is not enough. These actions, must also be accompanied by a new vision and new actions. And these actions need to recognize the challenges inherent in transformative projects and the situated realities of the people involved. If the collective human relationship with Earth cannot be sustained, then critique must be paired with a vision—and corresponding educational tools—that embrace the possibility of enacting a new relationship.[11]

Moving Forward

The first wild pedagogies colloquium was canoe-based and hosted on the Yukon River in 2014. This proved to be an important catalyst for the work of this book. It was premised on the idea that in order to engage with alternative relationships with place, and our own work, we needed to conduct ourselves differently. Only in this way could we gain access to thinking and being differently in the world. This initial effort revealed some interesting teachings. It is true that the group was immersed in a relatively wild place, but, in some ways, the colloquium was still structured like a traditional conference. In particular, presentations were organized in a more-or-less predetermined sequence, and in most instances, they were not particularly responsive to the places where they were delivered.

Sure, being on a riverbank or in a canoe added additional elements to the work—the content and the venue were better aligned. And, surely the river added inspiration to the reflections, but in most instances, actual voices from the land were not recognized in the presentations and did not significantly shape the nature of the colloquium. We could not honestly claim that place was a co-teacher or co-researcher. This realization provoked the question: Could a colloquium be structured in a way that "voices" from the land and its more-than-human beings were better heard, and then play an active role in informing and deepening discussions?

Deliberations about the past colloquium gave rise to four important themes in the planning the May 2017 *Wild Pedagogies: A Sailing Colloquium*, hosted off the West Coast of Scotland. First, it was important to improve on our manner of scholarly interactions as experienced in the previous colloquium. An aim was to get beyond the stiffness of individualized presentations. We wanted to find a shared project that would allow conversations and thinking to grow and build, and that could actively press back against the isolationist tendencies of scholarship. Then, if this colloquium was to result in a collaborative book project—how were we going to write it? What processes could lead the collaboration and consensus required to give the theme "wild pedagogies" some depth and coherence? Second, enabling new relations with place would need to begin by recognizing other-than-human agency in a similar sense to the notion of nature as co-teacher or co-researcher. This would require deliberate daily activities during the colloquium and a move away from the human-centredness of most scholarly activities. Achieving the second theme could be more easily achieved by, third, placing the colloquium in a relatively wild place, far beyond the typical conference centre. This could allow our group to immerse themselves in a landscape and alter the dynamics between humans and the rest of the world. Fourth, this colloquium needed to be located some place where participants could be inspired to interrogate their own cultural norms, as colloquium participants were all privileged professionals and scholars from industrialized western countries. Ideally this would be a place that had already undergone a critical conversation within the context of such norms.

Scotland: A Sailing Colloquium

Colloquium planning rested on the premise that the design and place of a wild pedagogies gathering would ultimately shape the nature of discussion. Mindful of this inevitability, the colloquium activities and the Scottish location were purposefully chosen. In important ways, the planning process began with the word "colloquium" itself, which was wonderfully described by Louise Profeit-Leblanc:

> I found it intriguing that a colloquium basically means "in and of talk"; in other words, familiar speech or "talk-speak" if we can coin it as that! A place where people come to learn by listening and speaking with each other about the subject at hand.[12]

As an Indigenous storyteller deeply connected with her oral traditions, it is fitting that Profeit-Leblanc should have expressed this so well. A colloquium is meant to be far more conversational than a conference. Here conversational meant listening carefully, responding with open minds, and understanding in good faith. It is also about learning together and, these days, this is often framed in terms of social learning—or even better, eco-social learning. Finally, Profeit-Leblanc locates conversation in a place. For our purposes—and we expect for her, too—choice of place is integral to the kinds of conversation that can occur. If we are to make a move towards eco-social learning, where more-than-human others have a place in our conversations, then it matters where we have them.

To meet the needs of conviviality and conversation amongst humans, and others, and to position the colloquium in a relatively wild place, we travelled through a landscape on the West Coast of Scotland on a large ship, *The Lady of Avenel*. During the week we participated in sailing the boat, being in land and seascapes, living and eating together, engaging in activities planned to further knowledge and connection, and following a facilitated plan that helped to frame colloquium discussions. The colloquium was, thus, designed as a "total immersion" experience that allowed opportunities to interact with the landscape and deepen the theories of wild pedagogies in a shared and organic way. While living on *The Lady of Avenel*, the landscape had a more active role and the sea was ever-present in the rocking motion, the creaking of the hull, and the sounds of wind in the rigging and sails.

> *From the Ship's Manifest: Wild Pedagogies Version*
>
> *May 8th, 2017*
> *Captivated by the activity of sailing.*
> *Setting and trimming the square sails.*
> *Climbing the mast. Sitting in the sun.*
> *But still, the wind shifted, telling us to descend.*
> *Before reaching the deck that square sail—the course—was billowing against the shrouds.*
>
> * * *
>
> *May 12th, 2017*
> *Sleek boat pushing through smooth seas, manifold greys in the sky, shifting dimensions in tone, wave, wind, cloud, islands. We sneak through a narrow passage in the midst of the Garvellachs. P- waxing lyrical about glacial till during lithification of seabed, which settled 935 million years ago and was later thrust up through the crust at oddly wave-like angles. The islands dance. The Lady of Avenel surges through the sea gracefully, so close to rock, shags, seals, and eider ducks.*

Image 1.1 Wind change in the rigging. Photo credit: Aage Jensen

Individual space on sailboats, even large ones, is always limited—someone is always nearby. This can be an ideal venue for conversation. Our design included opportunities for structured conversations and a great deal of time for reflection and more casual chats. Rather than having formal conference-style presentations, two facilitators were recruited to create activities and track questions that provoked thoughtful conversations throughout the day. They provided periodic syntheses of preceding conversations for the purposes of furthering the joint writing project that was to become this book. They also actively created space for on-going check-ins, programme adjustments, avenues for furthering important or incomplete elements of the syntheses, and finally they encouraged us all to be continuously more aware of the place itself. This allowed for themes and discussion points to arise not just from within the group of humans but from the landscape of our travels as well.

Travelling this way also contains design pitfalls. While there are always compatibility risks, we would like to focus on one of the items central to our project that required some intentionality. Here we are talking about how we set out to make the voices of the Scottish landscape more manifest—how we set out to actively listen with better attunement. Our hope was to see what we had not seen before and to allow space for more-than-human discussants to take the fore. All this, we believe, is a necessary prelude to recognizing nature as co-teacher and conversationalist.

Those with experience travelling in groups know that companionship plays an extraordinarily important role—so much so that it can easily displace experiences of being present in a place. Having fun together and engaging seriously with human counterparts is an important part of the experience, but wild pedagogies suggest there should be more to an educational experience. This phenomenon goes hand in hand with the threat of our own writing project becoming insular, and isolated from the landscape—despite our expressed interest in nature as co-teacher, and in collaborative scholarship. It is easy for us to slip into "living in our heads" while only focusing on the human. We think that being physically located in a landscape is important, but not enough.

Our response to this human-centred pitfall was to set up some concrete activities and to encourage spontaneous attention. First, cultural and natural historians were invited to join us. Their observations and insights drew our attention to the place and its beings. They participated fully in our discussions and were constantly available "on deck" for impromptu talks and casual conversations. Their presence and participation was an important aspect of our journey.

Second, participants were invited to take time to be on their own both on the ship and ashore. Sometimes this also included opportunities to write short vignettes to describe moments when their experiences on our physical journey resonated with moments in our wild pedagogies journey. Some of these are sprinkled throughout this text to collaborate with the wild pedagogies ideas, and to make present the landscape and its beings for the reader, and as active members of our conversation.

Third, and slightly more structured, we kept a ship's manifest of our own design. It functioned in some ways parallel to the ship's official log.

In our case we requested daily entries that made explicit the wild voices and presences that we encountered. Some of these entries, too, will be presented as vignettes when aligned with our reflections.

Finally, we encouraged disruptions. We often stopped our conversations, or had them stopped for us, in the presence of birds, blooming bluebells, and when joined by porpoises. We looked, listened, marvelled, and tried to make sense of the more-than-human voices we encountered. We paused, took deep breaths, and sometimes smiled. Though, to be honest, nothing unfolded as perfectly as it sounds when written here. Sometimes our agendas were at odds. Sometimes the intensity of our conversations turned inwards and overwhelmed the presence of more-than-human-others—but not always.

Collaborative writing was also embedded in the design of this project and included a writing retreat at the end of the boat journey. Everyone stayed on for at least three days on the small island of Luing, and some stayed for nearly a week. Academics often—though not always—withdraw from their colleagues to write, and we are typically evaluated on our individual contributions. This can be the case even in multiple authored papers and books, where distributed components of the product, and edited drafts shift electronically from office to office at something like the speed of light and often our individual fragments are, in the end, patched together as best they can without sustained efforts at a more ecological kind of thinking. And, we often write for each other, less often trying to bring our ideas to the general public. Even less often will a voice *from* (as opposed to *about*) the more-than-human world enter the narrative. In all of this there is, of course, implicit irony. As educators, we are more comfortable experimenting with our practices, trying to transcend conventions, and taking pedagogical risks. As writers, this seems harder. So, this proposal gave us permission to experiment with scholarly conventions.

The previous paragraphs have spoken to the intertwined way that three of our thematic aims were framed. We have touched on how we aimed to work together in ways that could take us outside of scholarly conventions, open up opportunities for meaningful interactions with the more-than-human world, and position ourselves in a relatively wild place, ever-further from conventional conference centres. In the decisions that were taken, we aspired to align the collaborative processes with the aims of the wild pedagogies project. Still, given our fourth conviction, that the specific location of this colloquium would shape the nature of discussions: Why Scotland?

Recognizing that we were culturally, economically, and educationally a fairly narrow band of humanity we took to heart the suggestions of anthropologist Deborah Bird Rose.[13] For Rose, cultural change can be supported through active engagement with and reorganization of one's narratives. She suggests that within any culture there are histories, ideas, memories, even myths that do not align with the dominant narrative. And these might become elements for beginning to tell a different story of one's culture. Thus, in Scotland we found intertwined narratives of colonial appropriation of land, and lingering traces of deep relationships to land and that, in the end, brought us to the crofters, and some sixth century monks. These histories proved to be good grist for our work.

Contemporary conceptions of wildness and wilderness are linked to a nexus of ideas about geography, natural history, and landscape intersecting with political history, particularly that of imperialism and its twin, colonialism. This brew has often provided fanciful and idealized versions of wilderness. Scottish writers have been actively engaged in a critique of wilderness for a long time. Indeed, in the 1950s the renowned ecologist Frank Fraser Darling coined the phrase "wet desert" to describe areas considered by many to be wilderness when, in fact, those same areas were characterised by intensive land management practices including deforestation, soil drainage, muirburn, and grazing.[14]

Scottish historian James Hunter wrote a powerful version of wilderness critique in *On the Other Side of Sorrow*.[15] In this book, he argues that Scotland was a training ground for developing British imperialistic practice. And, once Scotland was subdued—and cleared of many of its Indigenous residents—it was a character no less than Queen Victoria that was at the forefront of idealizing a pastoral view of the resulting "wilderness."[16] As it turns out Scotland has a story to tell. Hunter's work provided a counter-narrative within the cultural norms of the colloquium participants. And, his Scotland provided a place to reflect on this counter-narrative, complete with suggestions of historical locations that could be visited from the *Lady of Avenel*.

Image 1.2 The *Lady of Avenel* lying off on Iona. Photo credit: Bob Jickling

Some of Hunter's ideas will be discussed in more detail later in this book. But during the planning stages of the wild pedagogies colloquium, his book offered some concrete grounding in a particular place. First, Hunter builds on the critique of an out-dated conception of wilderness, as was often used in Scotland. What some saw as an idealized wilderness, Hunter saw as a tragic absence of people from their homelands.

Second, whilst others have critiqued the oft-prevailing visions of wilderness, Hunter proposes an antidote to these out-dated visions. For him this lies in inspiration taken from translations of ancient Gaelic nature poetry written in the sixth and seventh centuries. Much of this was composed on islands off the west coast of Scotland. Perhaps the most famous of these islands is Iona. Here, according to Hunter, the monastic community was one of the intellectual centres for all of Europe, and a centre for literacy. These Gaelic-speaking monks were also remarkably at home on lonely nearby islands, such as *Eileach an Naoimh* in the Garvellachs group. It was in such isolated retreats that much of the inspirational poetry was composed. Hunter contends that this body of ancient Gaelic poetry describes fundamentally different relationships that these monks had with their surroundings, especially when compared with those of most humans today. Interestingly, as centres of literacy, these monastic communities drew inspiration and insight from the places themselves; this was not a detached literacy. Accordingly, Hunter's antidote to an idealized conception of wilderness, most often portrayed as land empty of humans, was to look to examples of different kinds of relationships between people and the lands they inhabit. Thus, reading the poetry of these island monks, where relationships appeared immersive, attentive, and respectful, might provide some clues to a better approach for environmental change.

Third, many locations described by Hunter are on accessible islands in what are now known as the Inner Hebrides (*Na H-Eileanan A-Staigh*), off the west coast of Scotland. This meant that our conversations could take place on a large sailing boat while visiting sites of historical, natural, and cultural significance. We could literally visit ancient stone cells inhabited by the monks that are such an important subject of Hunter's interests. Then we could sit in front of these cells and read translations of the Gaelic poems amidst the calls of actual corncrakes and cuckoos, the great-grand offspring of those that filled the poets' lives.

Eilach an Naoimh, in the Garvellachs

We stepped onto Eilach an Naoimh in the Garvellachs group of islands. This rocky place is on the west coast of Scotland. The Irish monk Brendan is said to have preceded our visit by almost 1500 years in the year 524. Later in the 6th Century, St. Columba settled here, too, before moving on to found a monastic community on Iona. It isn't easy to get to this island. Anchorages are exposed to the weather; the shoreline is rough.

This island was a place of retreat. Here, monks lived a solitary and contemplative existence. They stayed in stone cells resembling beehives and constructed of expertly placed flat stones that spiralled upwards and inwards. It must have been a simple—and perhaps lonely—existence. Interestingly, their contemplations were rooted in the particularity of their place. And these, together with their observations, were poetically recorded in the Gaelic language.

It was a warm May day when visited Eilach an Naoimh. I sat on a flat stone in front of the remains of one of the beehive cells and read a translated, and now anonymous, poem written more than a thousand years ago:*

> May-time, fair season, perfect in its aspect; blackbirds sing a full song, if there be a scanty beam of day.
> Summer brings low the little stream, the swift herd makes for the water, the long hair of the heather spreads out, the weak white cotton-grass flourishes.
> The harp of the wood plays melody, its music brings perfect peace; colour has settled on every hill, haze on the lake of full water.
> The corncrake clacks, a strenuous bard; the high pure waterfall sings a greeting to the warm pool; rustling of rushes has come.

As I uttered its name, a corncrake began to call. We were struck dumb. For in this moment we shared an experience with the poet that stretched across time. We listened silently until it had finished, and then went on.

> Light swallows dart on high... The hardy cuckoo sings, the speckled fish leaps... The glory of great hills is unspoiled.

Suddenly, and again just as its name was mentioned, a cuckoo called out. We listened again, and this time the corners of my eyes began to water. When this voice went silent, I finished reading.

> Delightful is the season's splendour, winter's rough wind has gone; bright is every fertile wood, a joyful peace is summer.

Like the ancient poet-monk, we, too, experienced a joyful peace that afternoon in May.

*From James Hunter, On the Other Side of Sorrow, pp. 45–46.

Why Wild Pedagogies? 17

Image 1.3 Ancient stone cell on *Eilach an Naoimh*. Photo credit: Hansi Gelter

The landscape was important too. This part of Scotland is profoundly geologic, denuded as it is of most of the tree cover over thousands of years of human occupation. The geology is remarkable, going right back to pre-Cambrian times, with worn and ancient rock, over 935 million years old forming some islands. More recently, Cambrian, and later volcanic "intrusions," have created dramatic layers of lava and quartz crystal, slicing through the older rock. The Isle of Staffa is an example of perhaps the most astonishing geology of all. Its lava flows cooled and cracked forming into enormous—predominantly 6-sided—basalt columns that define the island and give dramatic shape to Fingal's Cave, and provide shelter, nesting sites, and access to rich food sources for myriad puffins, fulmars, and shags.

Thus, many of our conversations, and much learning together, took place on board the square-rigged brigantine, *The Lady of Avenel*. And, a good bit of this book was discussed—and indeed written—on board, amongst these islands, and later on the nearby island of Luing.

Whom are We Writing for?

This book aims to have broad public appeal. And it assumes that educational change will not arise from any particular location. Inspirational pedagogues are found throughout formal education and outside of it, too. So, we aim to meet people where this book finds them.

Education takes place at home, at work, and in community activities—with our children, our peers, our friends, and our neighbours. Education takes place in museums, aquariums, parks, playgrounds, summer camps, and social service agencies. And, of course, it takes place in schools, colleges and universities. There are educational steps that can be taken by parents, students, community educators, and teachers. There are also steps that can be taken by school principals, curriculum specialists, superintendents, academics, ministers of education, business leaders, policy makers, healthcare providers, and politicians. The time for this collective education action is now. It is critical to examine thoughtfully human activities on earth—our deepest assumptions, ideals, values, and worldviews. This is work for everyone with interests in education and who are called to wild ideas about pedagogy.

Given the breadth of the potential audience, this book will endeavour to present serious ideas in a way that has broad public appeal. In this way we are mindful of Canada's public philosopher, John Ralston Saul.[17] He reminds us that reform requires widespread philosophical understanding of the options available, and their implications. Too often, he has observed, important voices are absent from public debate because their proponents are caught up in a world of narrow specializations and impenetrable dialect. With these thoughts we have endeavoured to do better; our aim is to communicate effectively and maximize engagement with allies. Towards this end, we have attempted to write more clearly and more freely of scholarly conventions, in this introductory chapter, and in the last two chapters on "Touchstones" and the "Afterwords." We have written with a little more detail in chapters "On Wilderness," "On The Anthropocene," and "On Education."

It might be tempting to think of wild pedagogies and the touchstones we present in this book as a tight framework for the future, but that would not be correct. Or at least, this is not our intention. We hope, rather, that this work will be seen as a heuristic. These two terms, framework and heuristic, are different in important ways. Frameworks provide more concrete visions about how things are, how they should be, or roadmaps for getting to a new place. As such, they assert more control over analysis and can be more prescriptive in their aims. But heuristics are typically defined as agents in the process of discovery. They can act as aids to understanding or even shortcuts into the work itself. They are provocateurs at the intersection of imagination and praxis. Their aims are more expressive and generative—more attuned to the *wild* reader.

In truth, as we move between geological epochs—between the Holocene and the Anthropocene—we are traversing new terrain. Humans have never before witnessed this kind of epochal shift or had to accept this scale of responsibility. No one knows what will happen or how we will need to respond; uncertainty is part of today's reality. We do hope that the generative intentions of this book will inspire responses that are imaginative, creative, courageous, and radical—because this is what our times require.

The Shape of the Book

Wild pedagogies arose out of a convergence of ideas about wilderness, education, and the emerging realities of a new geological epoch, the Anthropocene. In the cases of wilderness and education, this work is at least in part about reclaiming language and ideas that have been put aside and largely discounted for too long. That we are shifting towards something called the Anthropocene is a relatively new idea, but one with grave implications for Earth.

In a way, we begin by heading down three separate pathways. However, it quickly becomes apparent that there are important points of resonance between these ideas—or, convergences in these pathways. To begin, our first pathway takes us through discussions of wilderness itself. We explore real places and existential experiences with these places. We hold wilderness valuable, and see a need to refresh and reclaim an understanding of the central idea of wildness as uncontrolled—even free.

On a different path there is the seemingly ubiquitous presence of control throughout much of educational conversation and practice. While travelling on this track, recognizing the gates along the way, we wonder about possibilities for a wilder pedagogy, loosed from domesticating—as in taming—forces. The third pathway for our enquiry is the one we seem to be travelling as a species, engulfing everything else in our wake. Here we are talking about the Anthropocene, where the defining epochal characteristics are human induced.

These three introductory pathways are evocative, to say the least, and serve to provide context for our response, and are the basis for Chaps. 2, 3, and 4. In turn, these chapters develop our ideas about wilderness, the Anthropocene, and education, and examine their points of resonance—the interplay between wildness, education, and control,—particularly as they relate to an emerging conception of wild pedagogies.

It is one thing to talk about ever-more wild pedagogies, but it is quite another thing to implement these ideas. Thus, the fifth chapter, "Six Touchstones for a Wild Pedagogy," is framed as a practical guide to help educators think through their actions on the ground as they work at shifting their practices.

The final chapter, "Afterwords," is a collection of reflections on how individual educators see adapting wild pedagogies for their own educational roles, in their own places. The point is that what we have done needs to be seen as consistently troubling the idea of control. We do not want six touchstones to be seen as a rigid framework, but rather, as an agent of continued discovery.

Acknowledgements *Crex crex* is the taxonomical name given to the Corncrake. We have chosen this bird to represent our collective because it was an important collaborator in this project and because its onomatopoeic name beautifully mirrors its call—a raspy crex crex. For some reason, it chooses to fly over England and breeds in Scotland and Ireland. Presumably this is due to loss of habitat in modern England, but perhaps these birds sense some epicenter of empire there? Who is to know?

Notes

1. Here we are riffing off of: Bruno Latour, "Will Non-Humans Be Saved? An Argument in Ecotheology," *Journal of the Royal Anthropological Institute* 15 (2008): 459–475.
2. Kris Gutiérrez, "Designing Resilient Ecologies: Social Design Experiments and a New Social Imagination," *Educational Researcher* 45, no. 3 (2016): 187.
3. Ibid., 192.
4. Irina Bokova, "Foreword," in *Education for People and Planet: Creating Sustainable Futures for All*, ed. UNESCO (Paris: UNESCO, 2016), 5.
5. Bokova, *Education for People and Planet*, 2016, 5.
6. Zygmunt Bauman, *Liquid Life* (Cambridge: Polity Press, 2005), 12.
7. Ibid., 12.
8. Ibid., 12.
9. Stephen Sterling, "Assuming the Future: Repurposing Education in a Volatile Age," in *Post-Sustainability and Environmental Education: Remaking Education for the Future*, ed. Bob Jickling and Stephen Sterling (London: Palgrave Macmillan, 2017): 31–45.
10. "Sustainable Development Goals: 17 Goals to Transform Our World. Goal 4: Ensure Inclusive and Quality Education for All and Promote Lifelong Learning," United Nations. Accessed February 5, 2018, http://www.un.org/sustainabledevelopment/education/.
11. See, for example: Sean Blenkinsop and Marcus Morse, "Saying Yes to Life: The Search for the Rebel Teacher," in *Post-Sustainability and Environmental Education: Remaking Education for the Future*, ed. Bob Jickling and Stephen Sterling (London: Palgrave Macmillan, 2017): 49–61.
12. Louise Profeit-LeBlanc, "Transferring Wisdom Through Storytelling," in *A Colloquium on Environment, Ethics, and Education*, ed. B. Jickling (Whitehorse: Yukon College, 1996): 14–19.
13. Deborah Bird Rose, "Connectivity Thinking, Animism, and the Pursuit of Liveliness," *Educational Theory* 67, no. 4 (2018): 491–508.
14. Darling, F. Fraser (Editor) *West Highland Survey: An Essay in Human Ecology* (London: Oxford University Press, 1955).
15. James Hunter, *On the Other Side of Sorrow: Nature and People in the Scottish Highlands* (Edinburgh: Birlinn Limited, 1995).

16. See, for example: T.C. Smout, *A Century of the Scottish People 1830–1950* (London: Fontana Press, 1987).
17. John Ralston Saul, *The Unconscious Civilization* (Concord, ON: Anansi, 1995): 161–162.

References

Bauman, Z. *Liquid Life*. Cambridge: Polity Press, 2005.
Blenkinsop, S., and M. Morse. "Saying Yes to Life: The Search for the Rebel Teacher." In *Post-Sustainability and Environmental Education: Remaking Education for the Future*, ed. Bob Jickling and Stephen Sterling, 49–61. London: Palgrave Macmillan, 2017.
Bokova, I. "Foreword." In *Education for People and Planet: Creating Sustainable Futures for All*, ed. UNESCO. Paris: UNESCO, 2016.
Fraser, D., ed. *West Highland Survey: An Essay in Human Ecology*. London: Oxford University Press, 1955.
Gutiérrez, K. "Designing Resilient Ecologies: Social Design Experiments and a New Social Imagination." *Educational Researcher* 45, no. 3 (2016): 187–196.
Hunter, J. *On the Other Side of Sorrow: Nature and People in the Scottish Highlands*. Edinburgh: Birlinn Limited, 1995.
Latour, B. Will Non-Humans Be Saved? An Argument in Ecotheology. *Journal of the Royal Anthropological Institute* 15 (2008): 459–475.
Profeit-LeBlanc, L. Transferring Wisdom Through Storytelling. In *A Colloquium on Environment, Ethics, and Education*, ed. B. Jickling, 14–19. Whitehorse: Yukon College, 1996.
Rose, D.B. Connectivity Thinking, Animism, and the Pursuit of Liveliness. *Educational Theory* 67, no. 4 (2018): 491–508.
Saul, J.R. *The Unconscious Civilization*. Concord: Anansi, 1995.
Smout, T.C. *A Century of the Scottish People 1830–1950*. London: Fontana Press, 1987.
Sterling, S. Assuming the Future: Repurposing Education in a Volatile Age. In *Post-Sustainability and Environmental Education: Remaking Education for the Future*, ed. R. Jickling and S. Sterling, 31–45. London: Palgrave Macmillan, 2017.
"Sustainable Development Goals: 17 Goals to Transform Our World. Goal 4: Ensure Inclusive and Quality Education for All and Promote Lifelong Learning." United Nations, n.d. Accessed February 5, 2018. http://www.un.org/sustainabledevelopment/education/.

freedoms and abilities to live and dwell on-their-own-terms. What seems clear is that wilderness, framed this way, is descriptive of real places characterized by their inherent wildness. Wilderness, in this case, is not simply reduced to a human idea. As such, wild places can be characterized in part by an emergent will of its inhabitants to realize their own ends. Or, to put it another way, wilderness can be thought of as a place where there is freedom to flourish. These characterizations are, for us, essential components of a re-thought and re-negotiated wilderness.

> ### Cnoc Buidhe (Yellow Hill)
>
> Our dingy landed on a white sandy beach on the north-easterly shore of Iona. My first destination was further north yet, to the last croft on the Island—Lagandorain. I'd first come here about fifteen years earlier, just a few years after John MacLean acquired the land.
>
> John was by trade an antique rug restorer from Edinburgh. Being a MacLean, his family comes from Mull and he can trace his ancestry a long way back. Moving to Iona had long been a dream: it was a formative place for him. He had a deep-rooted personal connection and knew it well. He also completed a poetic circle by returning to the land of his ancestors.
>
> His plan was to farm the croft and to supplement his income by running a small hostel on the property. However, this plan differed from the intensive sheep farming of the previous crofters. He was to raise a smaller "Hebridean" sheep thought to have been introduced by the Norse, a thousand years ago. More suitable to conservation, these sheep would reduce the grazing pressure and meet John's needs for a niche market. His plan was to craft artisanal rugs from the black wool and market them as a product of the island.
>
> Being by heart a restorer, John decided to fence off a few acres of land as a conservation area. This was essentially a little hill behind his house. As is often the custom in ancient lands, even little hills are named. This one was called, in Gaelic, Cnoc Buidhe. Translated, this literally means a small yellow hill. Yet, its yellowness had not been seen for some time. Meaning had been disconnected from memory. Dismembered.
>
> Shortly before my first arrival at this place, John was rewarded for his land stewardship. The grazing had been lifted and, as if from nowhere, the resting hill blossomed with yellow flowers. This is a story he told with a measure of wonderment.
>
> Indeed, this is something to wonder about. These flowering yellow plants had been subjected to generations of intensive grazing pressure. Their will had been literally clipped before it could rise to its blossoming potential— year after year. Suddenly it was there again, its will—its drive to bloom— was realized. Its wildness was again manifest.
>
> Upon our arrival this year, the hill looked lush. At its base was the abundant foliage of yellow flag iris, on the brink of blooming. A little further up the

> hill, we could see a few blooming primrose—yellow of course. In the intervening years John has learned more. Cnoc Buidhe is more than a yellow hill. It is place for primrose, celandine, yellow flag, marsh marigold, birdsfoot trefoil, bog asphodel—all have yellow flowers. As spring moves through summer and into fall even the sphagnum and grass take on a yellow hue.

Image 2.1 *Cnoc Buidhe* (Yellow Hill). Photo credit: Bob Jickling

Engaging with wilderness in wild pedagogies is intended to cultivate an ability to recognize the significance of real—material—places and the freedoms of all life to flourish, human and other-than-human. When these ideas are placed in contrast to ideas and ways that tend towards control, they have the capacity to disrupt common thinking and practice. Having revisited the etymological roots, we will continue re-thinking wilderness by considering other examples of how this idea has been conceived and deployed historically. We will also consider the language we might use to revitalize wilderness in the Anthropocene and how this language can inform our teaching and environmental practices.

Colonization, Indigeneity and the Wild

There are many ways of looking at wilderness. Not surprisingly, Indigenous peoples around the world have mixed views about the idea—including raw hostility. "Protecting wilderness" has historically gone and continues to go hand in hand with disenfranchisement and marginalization. Access to traditional territories, homelands, and cultural continuity is disrupted—often violently—through predominantly colonial evocations of "protecting wilderness" and idealized visions of an unpeopled paradise.

In the book *On the Other Side of Sorrow*, James Hunter describes how the colonization process unfolded in the Highlands of Scotland in the eighteenth and early nineteenth centuries. With rising industrialization and textile manufacturing in England, it became apparent that the hills and glens of Scotland could, if cleared of the human inhabitants, be more profitably used for large-scale sheep farming. In concert, he argues, the patriarchal rulers of the family-based clan system degenerated into a class of rapacious landlords. The tenant farmers, or crofters as they were called in the Highlands, were not profitable; they were in the way; their culture and way of being on the land was no longer welcome.

The ensuing Highland Clearances—the forced and often violent removal of people from their homelands—has been extensively analysed. This clearing of people was achieved through inter-clan fighting, dramatic rent increases, intimidation and violence, and an assault on traditional Gaelic culture. Moreover, it was systematically assisted through a process of land commodification made possible by new forms of economic expediency that benefitted those who controlled land use. For Hunter, all this was underpinned by:

> a profound conviction, held in Edinburgh every bit as much as it was held in London, that the people thus being brought to heel were an irredeemably inferior set of human beings.[9]

According to Hunter, it was in the Highlands of Scotland that the British Empire honed its imperialistic capacities.[10] Though, surely these capacities were also developed in other parts of Britain as well.

By the 1840s, when Queen Victoria ventured into the Highlands, Hunter reports that she found the area "most delightful, most romantic;" and characterised the landscape by its: "quiet," "retirement," "wildness," liberty," and "solitude."[11] Solitude, indeed. This, of course, had only been achieved by evicting, or aggressively encouraging resettlement of the Indigenous Highlanders, and stripping them of their cultural connections with the land. No mention is made in Queen Victoria's journals of the poverty and hunger experienced by the remaining crofters. Hunter claims, too, that Victoria must have known that renewed clearances were removing yet more Highlanders. Still, her wilderness, that "seemed to breathe freedom and peace," was a land of emptiness and sorrow for Scots. This point is punctuated by Hunter's account of a Highlander by descent that returned to deserted glens of his homeland, three generations later. Overwhelmed by the experience, this fellow could not help but observe: "everyone who ever mattered is dead and gone."[12]

Not surprisingly, the same process was repeated elsewhere. In 1872, the year marking the creation of the world's first national park (Yellowstone in the United States), several tribes of Indigenous peoples were forcibly removed from the park's designated land in the name of wilderness protection.[13] Wilderness was again equated with an absence of human inhabitants in the name of "protecting the pristine." Similar examples can be found around the world, right up to today. Given such examples, one can easily recognize the value and necessity of Cronon's call to re-think notions of pristine wilderness and the role such visions have played in colonization.

Our suggestion here is that one can agree with Cronon that characterizing wilderness as an absence of people is a problem without taking the further step of agreeing that wilderness is simply a human creation, and in need of discarding altogether. As such, we wonder if dismissal of wilderness as a material reality does justice to the complexity of a re-conceived concept of wilderness and the range of work that it might do. An example from Canada's Yukon Territory can, perhaps, open some fresh possibilities for reimagining a more interesting, just, and ecocritical approach to wilderness.

In 1993, a unique conference on "Northern Protected Areas and Wilderness" was hosted in Whitehorse. At that time, the peoples of the Yukon were immersed in complex negotiations aiming to establish a framework for settling outstanding land claims between Yukon First Nations and the governments of the Yukon and Canada. Indigenous peoples had not ceded any claims to the land, yet had suffered much during the process of colonial expansion. To make matters worse, some had been dispossessed during the establishment of a national park reserve in the Kluane area. The future of land access and rights to practise traditional activities was at stake. "Wilderness" was not a popular word for First Nations. Other, newer residents were frequently reminded that there were no words for "wilderness" in various Indigenous languages.

The rejection of wilderness was encapsulated by Chief Joe Johnson of the Kluane First Nation:

> I don't know where the word wilderness is coming from. Why is it wilderness when I feel comfortable when I go back there? What segment of the population call this area wilderness when it's not to me and my people?[14]

While these sentiments were common, Norma Kassi, a member if the Vuntut Gwitchin First Nation, reached out in a different direction. On a number of occasions she affirmed that there was no word for wilderness in her language, however, she did say that there was a related word for freedom.[15] In this offering she implied that perhaps the concept of wilderness might be linked to freedom. This is neither the hedonistic freedom of individualism, nor the capitalist freedom of "free trade." Hints to its meaning may lie in the hurt she expressed when humans "manage and study" animals. For example, she spoke about a caribou that her brother once killed that had been fitted with a radio collar. "Under the collar," she said, "was covered in worms, it was tight. I do not know how the caribou lived, it was skinny and segregated from the others."[16]

It is difficult for an outsider to fully appreciate the nuanced meanings of this Gwitchin word that Kassi likened to freedom. However, she used it purposefully, and in a way that resonates with the Old English "wildoerness" in that it reflects the wildness of the animals in her homeland. The point here is not to speak for Norma Kassi; rather, we cite this example for two reasons. First, in this reaching out, Kassi suggests that there can be culturally grounded ways of thinking about wilderness that do not demand the absence of people. Second, it is an offering from one culture to another in a gesture aimed at promoting understanding between peoples. And, in this case with the caribou, Kassi is hinting at ways that she, and others in her culture, can live in profound relationships with animals in their shared territories.

Similarly, James Hunter also looks for a culturally grounded alternative to the Victorian conception of wilderness that manufactured "freedom and peace" by removing most of the people, and dehumanizing those that remained. In his book he is at once, critiquing a destructive conception of wilderness, but also attempting to lift up a dispirited people through a pride that can be found buried deeply in their cultural history. Here he draws on the literary traditions of Gaelic poetry, traced back to sixth century monks. What also matters in their body of poetry is "the enormously sympathetic tone which it adopts when dealing with wild nature."[17] Hunter concludes that:

> these determinedly ascetic individuals, as readily apparent from their writings, developed what was—from a European standpoint anyway—a wholly new relationship with nature.[18]

Like Kassi, Hunter offers a culturally grounded way to think about wild places. Here an absence of people is not the critical component. Rather, living well means living within a very different relationship with nature—a relationship that includes deep immersion, careful listening, and respect for the other-than-humans inhabiting their shared communities. And again, maybe a concept of wilderness that embraces these qualities can do good work for wild pedagogies going forward.

As a short addendum to this section we acknowledge that we have used Scottish and Canadian examples of colonization. Other examples could be drawn from around the world. Hunter, for instance, deploys examples from Africa and North America in his book. Within our own colloquium group, two Nordic colleagues immediately raised the relevance of such conversations to their respective countries and interpretations of wilderness as they relate to the Sami people.[19] Their work, included as an endnote, serves as an example of how the ideas presented in this book here can be interpreted in light of regional differences and the specificities of particular places. We urge readers to consider and examine examples from their own regions.

Wilderness and Civilization

Marie Battiste, a Mi'kmaq scholar, reminds us that in a globalized world, humans are constantly "marinated" in a Eurocentric consciousness.[20] This consciousness typically holds a separation between humans and the rest of the world. Critiques of this separation often frame, as a key problem, the tendency to perceive the world to be made up of habitual dualisms such as: subjects and objects, humans and non-humans, men and women, colonizers and colonized, culture and nature, civilization and wilderness, modern and primitive.[21]

Typically these dualisms come to be understood as hierarchically related and are reinforced in language, social practices, curricular documents, learning objectives and in the material configurations of our learning environments. Often the assumptions upon which these dualisms rest are buried so deeply within social psyches as to be functionally invisible. However, a key assumption that can be teased out of the examples given is that humans are positioned at the centre of everything, further implying that they are also at the top of a paired hierarchy, and are thereby most important. Establishing new relationships will require disrupting assumptions of both hierarchies of importance and centrality,[22] and associated root metaphors.[23] A task for wild pedagogies is to make such habits of mind explicit and available for critical analysis.

One of the issues we take with Cronon, is that his critique of wilderness does little to disrupt human centredness. By declaring that wilderness is solely a human construction, he attributes an amazing amount of control to human beings—that we control our own narrative interpretation of ourselves and that this interpretation has the ability to shape the world around us. For such a thesis to work, the natural world and all of its beings, entities, weather patterns, and relationships must be understood as essentially un-involved, inactive, and of no consequence.

Elsewhere, Cronon suggests, "We *could* choose to think about nature differently"[24] and claims that humans can alter conceptions of nature at will without input from the natural world. It follows from this that nothing in nature has agency and that nothing in nature plays any role in shaping how humans are constituted or in how they understand wilderness. Thus, for Cronon, even if humans choose to think about wilderness differently the result would still be a cultural construction that reflects human judgments, values, and choices alone. Unfortunately, such logics reinforce narratives of human centrality and control rather than questioning our place in the world or in recognizing our inherent connections to more-than-humans.

To be fair, human desires, values, and judgements do play a role in shaping how we perceive the world and it is important for us to attend to this power, possibility, and danger. However, wild pedagogies seek to challenge a human-centred, and human exclusive, understanding of wilderness. More complete understandings of wilderness will be those that honour the voice and presence of nature and more-than-beings.

Towards the end of his paper, perhaps sensing the potential for misinterpretation, Cronon backs away from his insistence that humans alone construct wilderness and nature. This oft-overlooked portion of his essay may provide some possible inspiration for wild pedagogues, or at least a response to those who have repurposed his thesis:

> The romantic legacy means that wilderness is more a state of mind than a fact of nature, and the state of mind that today most defines wilderness is wonder. The striking power of the wild is that wonder in the face of it requires no act of will, but forces itself upon us—as an expression of the nonhuman world experienced through the lens of our cultural history—as proof that ours is not the only presence in the universe.

> Wilderness gets us into trouble only if we imagine that this experience of wonder and otherness is limited to the remote corners of the planet, or that it somehow depends on pristine landscapes we ourselves do not inhabit. Nothing could be more misleading. The tree in the garden is in reality no less other, no less worth of our wonder and respect, than the tree in an ancient forest that has never known an ax or a saw—even though the tree in the forest reflects a more intricate web of ecological relationships.[25]

Similarly, in the foreword to a newer edition of the book containing his controversial essay, Cronon attempts to clarify his position, noting the danger of conflating conceptual critiques with material realities:

> Asserting that "nature" is an idea is far from saying that it is only an idea, that there is no concrete referent out there in the world for the many human meanings we attach to the word "nature." There are very real material constraints on our ideas and actions, and if we fail to take these into account, we are doomed to frustration if not outright failure. The material nature we inhabit and the ideal nature we carry in our heads exist always in complex relationship with each other, and we will misunderstand both ourselves and the world if we fail to explore that relationship in all its rich and contradictory complexity.[26]

These questions regarding the material and conceptual are important themes that we will return to later in this chapter.

Language and Ecology

An important insight from the preceding sections on wilderness points to the power, but also the difficulty of language. What language will be required in wild pedagogies, for example, to avoid falling back into dualisms between nature and culture or civilization and wilderness? Deeply held separations—and their capacity for bending all things cultural back towards a *status quo*—seem inevitable unless we can enact a more humble humanity. This requires working towards understanding our overlapping relationships and meaning-making with a reciprocally engaging world. Indeed, concepts such as wilderness may eventually no longer even be required not because Cronon was right but because Chief Joe Johnson was. We cannot dissolve thousands of years of boundary-making with an intellectual assertion, as in "nature is a human construct" or "we are all nature." New relationships and commensurate language will arise slowly out of action—actual engagement in new ways of being present to, and interacting with, the world.

Perhaps a first step in shifting language and challenging human dominion is to disrupt taken for granted patterns of usage. David Abram,[27] did this when he rejected the term "non-human." There is a tendency in the English language to speak of humans and then represent everything else by the judgmental prefix "non." His response was to coin the term "more-than-human," that was meant to diminish the status of humans both ethically and in terms of significance by pointing to the "moreness" of Earth. As disruptive as this move has been, it isn't perfect. It still reduces all other beings into a single descriptive category.

Others have attempted similar disruptions in order to decentre the human including the "relative wild"[28] and "ecology without nature."[29] Perhaps more evocatively, Some Indigenous cultures use a phrase like "all our relations."[30] Still, what would it take for current non-Indigenous cultures to live in ways that words like these could be genuine, and not just co-opted?

Val Plumwood[31] takes a different approach in a chapter appropriately called "Paths Beyond Human-Centredness." She begins by reminding us that human-centredness causes us "to lose touch with ourselves as natural beings, embedded in a biosphere, and with the same dependence on a healthy biosphere as other forms of life."[32] In developing her arguments, she suggests that *prudence*—that is, taking care of ourselves—always demands a certain degree of human-centredness.[33] However, she also argues that it is possible to consider our own interests *and* the interests of others. Why then, she posits, should we have different standards for other-than-human cases? Given the inevitability of "centrings" she calls for positive and multiple centrings.

In this same spirit, some of the wild pedagogues in the present project came up with the idea of a "more-than-plankton world" as a way to decentre humans and evoke the interconnected but distinctly separate species, rock strata, ocean currents, and lives that makes up this planet. Rather than becoming simply another hierarchy, the "more-than-plankton" notion can also remind us that each species has its own locus of meaning, and "interprets" the wider planet from its own position.

As we have attempted to show, it can be difficult work challenging and moving beyond cherished concepts, metaphors, and dualisms. The discussion in this section has tried to capture the difficulties in talking about our experiences in a living world. An important task for wild pedagogies will be to recognize and acknowledge these experiences, and represent them without falling back into human-centred ordering. A parallel task will be to challenge damaging concepts, metaphors, and dualisms when we encounter them. However, it will not be enough to just disrupt the root metaphors embedded in common language use. Our task, as wild pedagogues, will also require recognizing that this multispecies, multi-centric world is a communicator. We might work to help interpret what we hear and also to help find ways to frame these messages from the world in more just and respectful language.

Learning to Listen in a More-Than-Human World

While it is important for humans to re-think language and metaphors, engaging with the more-than-human world with reciprocity will require more—it cannot be passive. Indeed, as we argue throughout this book, this will mean engaging with "somethings" and "someones" beyond us. For this to occur, we recognize that wilderness has *agency* or as many wild pedagogues like to say, *voice*.

What do we mean when we say listening to the *voices* of other-than-human beings? Everyone knows only humans have speaking voices; other things have yelps, chirps, whirls, whooshes, and hums perhaps—but surely not voices? After all, conversing with animals or inanimate objects has long been a hallmark of madness in "civilized society." And yet, we might say, learning to attend to the voice of more-than-human others lies at the very heart of wild pedagogies. But what could trilling shorebirds and the rhythmic swell of the sea be saying? What could beach rocks and the barren crags of seaside cliffs possibly teach us?

The notion that nature sometimes "speaks to us" is common enough, but the dominant mind often assumes this expression to be metaphoric. The sea did not actually "speak," you just felt something inside yourself and attributed it to the experience of standing on the shore. We readily accept that whales sing, seals roar, and the wind whispers, but there is a sense that these are again metaphors, not statements of fact. A literal interpretation would be considered unscientific, sentimental, or just plain weird in most staff lounges and high school classrooms.

Wild pedagogy, on the other hand, begins with the assumption that other-than-human beings and Earthly processes literally have voices.[34] Birdsong tends to be the most obvious and defensible example. Clearly birds have what might be called "voices" and it is not uncommon for people to actively listen, interpret, and often reciprocate calls with them. But can voice be understood beyond sonic expressions discernable to the limited senses of human beings? There are both cultural and educational precedents that suggest the answer could be yes. For example Abram points out that for most cultures, including our own in preliterate times, the entire world spoke; the birds, rain, trees and the land all possessed voice.[35] Similarly, place-based educator David Greenwood has written: "A theory of place that is concerned with the quality of human world relationships must first acknowledge that places themselves have something to say. Human beings, in other words, must learn to listen (and otherwise perceive)."[36]

One way of looking at voice is to recognize that every entity and every Earthly process sends out signals: as movements, activities, sound waves, radiation, or as reflections of sunlight. These signals are some of the voices of nature. The rhythms of ocean waves, the migration of species, the seasons, the clouds, changes in temperature, the growth of trees, all the fractal structures and patterns of nature can be considered the voices of other-than-human beings. Historically, these voices of nature were critically important for human survival. Elders, children, shamans, Sami noids, and others were interpreters who brought these voices into the social discourse.

Learning to listen to such voices, however, requires time and openness. In human terms, if you are not exposed to the practices and the experiences of being in a foreign culture, it is difficult to understand the meaning of the language. Picture yourself. You have just arrived. People call to one another, engage in discussion, messages transferred, decisions made, actions taken. You know this is going on because you are witness to it, but you do not understand much of the hows, whys, or wherefores. Similarly, in a forest you may notice the beauty of birdsong, you may even rest to enjoy the range of voices around you, but soon you continue on your way without having understood much.

We suggest that one of the major educational barriers to recognizing the significance of more-than-human voices is simply not having opportunities and time to cultivate the attention required. Too often in the modern world we meet other people, or students, that demonstrate a troubling deafness to the world. One member of our colloquium recalled a trip to Antarctica, noting fellow researchers were clearly not heeding warning signals given by sea lions and sea elephants, and then they were taken off guard when the animals attacked. Another member of our group offered an example of a group of environmental educators canoeing in the wilderness of northern Canada that did not attend to the frantic voices of a pair of loons. Unaware, or misunderstanding the message, they pitched their tent next to a small pond where the loons had just laid their eggs.

We are all, to some extent, enclosed in various bubbles that filter or mute signals from the world. We have technological bubbles such as the endless distractions of smartphones and computer screens. And we can find ourselves in social bubbles when we become absorbed by our group's interests and discussions. Embracing wilderness means, to some extent, breaking free of these bubbles—at least some of the time—so that we are able to attend to the world beyond narrow personal and human-centric concerns. A part of wild pedagogies can be "walking the talk" and fostering practices and orientations to recognize voices of the more-than-human world, and their educative significance.

Reverberating the Wild

This chapter has set out to re-think wilderness. We began with bold contrasts, an attempt to undo the human desire to be the centre of everything, and in control. And we have pushed against habitual dualisms. We have suggested that wilderness might be characterised as being self-willed, and places where there is freedom to flourish. This seems like a reasonable starting place to outline new components of wilderness. If we are, indeed, entering a new geological epoch commonly referred to as the Anthropocene, then the things that humans have excelled at are colonization, domination, and destruction. And, humans are constantly being marinated within these "talents." Strong ideas are needed to rattle human *hubris*, and to create openings for different possibilities.

In concluding his essay, Cronon, too, reached out to offer important components for a rethought wilderness. For him the power of the wild lies in the wonder it generates. This is not a passive wonder, rather it is wonder that forces itself upon us. Wilderness, here, has agency. In this, Val Plumwood helps us move forward. She speaks about how human-centred frameworks prevent us from experiencing others in nature in their fullness. When this happens,

> we not only imperil ourselves through loss of sensitivity but also deprive ourselves of the unique kind of richness and joy the encounter with the more-than-human presences of nature can offer. To realize this potential, we will need to reconceive the human self in more mutualistic terms, as a self-in-relationship with nature, formed not in the drive for mastery and control of the other but in a balance of mutual transformation and negotiation.[37]

As Plumwood reminds us, we live in on-going relationship with a larger matrix of beings. We may imagine that we can control this matrix, but if we humans attempt to do so, we will be continuously surprised at how unexpected the various beings and processes really are.

Humans who care to look will find that this complex matrix of beings and elements is not simply an idealistic absence of intrinsic controls. We see this in the repetition and stability of form: in ripples and waves, in spring migrations, even in the systole and diastole of blood circulating through our own veins. Challenging the idea of control does not mean all control is dismissed or that "anything goes." Rather, it entails thinking more carefully about what constitutes destructive controlling and what essentially healthful controls might look like. Having a heart that pumps in response to the bodies' need is the kind of control that allows for mutual benefit.

Living beings are also homemakers, "domestic" in their own right. Take the puffins that we saw nesting on the rocky cliffs of Staffa, off the west coast of Scotland. They nest and rear young in a niche—a stable set of repeating relations—and they are active participants in the negotiation and transformations within this niche. In their homemaking, puffins make a living, tend to their families, and reduce danger. It is out of this participation in this niche that puffins' unique way of being in the world emerges. They are puffins in a more-than-puffin world. There are, we think, some insights here that can help wild pedagogues to guide us all in the direction of more graceful relationships within our larger shared home—Earth.

Puffins

Atlantic puffins are always a highlight for me when I visit Staffa. Of course there is the famous cave and the dramatic geology of the place. But with their smart, tailored coats and brightly coloured bills, puffins rule the roost. I appreciate how they are often carrying out their business only a few feet away from clicking shutters, and yet they make time to oblige the day-trippers from Oban or Tobermory with their photogenic profile. Most see them as charming, almost comical actors amongst the sea pinks and tussock grasses. This is where they make their burrows, court, and defend their nests and bring bills full of silver sand eels to their young pufflings during the summer breeding season.

But for me, knowing something about the birds—where they go when they are not breeding on Staffa's green cliff top turf, what they eat, and how they get their food—brings a deeper perspective and richer respect for puffins and their relationship to place. Some of these birds, I know from ringing (banding) studies, can live into their mid-thirties—older than many of the researchers studying them. And I wonder what a 30-year-old puffin's view of Staffa and the surrounding seas might be?

> *We also know that puffins spend much of the non-breeding season (that's 7 months out of 12—the majority of the year) in the mid-Atlantic Ocean. Changing ocean currents, changing plankton populations, and hence fish and competitor seabird numbers, can impact where and when it is possible to survive the rigours of an Atlantic winter. I am amazed by their resilience and often wonder how a puffin knows where and when to head to a particular patch of sea where the food is? What information is it gathering from other puffins, from the weather and larger climate patterns, from the temperature and salinity of the seas it is diving in, or from the movement of other seabirds and perhaps whales? How does it learn these things? Who are its teachers? But more than this, simply in observing the puffins of Staffa, their remarkable beauty, resilience in the face of an extreme environment, their ability to pull their fishy food out of a seemingly vast and empty ocean, I find a different perspective and a new understanding of nature and our place in it.*

Image 2.2 A Puffin on the Island of Staffa. Photo credit: Hansi Gelter

Like control, the notion of freedom is not an unrestrained idea. It is routinely utilized to justify global "free markets" without restraint of regulation. And it is used to justify "free choice" in schools without regard for marginalized families that have fewer choices available to them. But, freedom can also be understood as something exercised by individuals existing within a community in a network of co-evolutionary relationships. Here, freedom comes coupled with responsibility. The cells in our bodies require fresh in-flows of blood constantly and, were the heart to freely choose to pump based on its own vagaries, the body and then ultimately itself would die. We can rightly recoil at the knowledge that species are becoming extinct at alarming rates through no fault of their own. Their freedom to flourish—and even exist—is denied by free choices and acts of control made by others who are ignoring their responsibilities. As Kassi was pointing out with the caribou example, we can only really achieve freedom when others, too, have freedom to flourish.

In a final point of clarification, we return again to the concluding portions of Cronon's essay.[38] He reminds us that talk of wilderness goes awry when we imagine it only occurring in remote corners of Earth, pristine landscapes, and uninhabited places. Wildness he asserts, occurs where we live: "The tree in the garden is in reality no less other, no less worth of our wonder and respect, than the tree in an ancient forest that has never known an ax or a saw." This is consistent with the re-thinking we have been doing in this book. Wildness exists in many places, however industrial or urban. This is important to acknowledge for various reasons. Importantly, it means that we can find access to the wonder of wildness in our home landscape.

It is important to notice, however, that Cronon begins by talking about wrongheaded notions of wilderness but goes on to talk about *wildness*. These terms are not the same. Wildness can reverberate in woodlots, parks, school grounds, and vacant lots.[39] It may even reverberate within our own classrooms.[40] While these places can be a source of inspiration and wonder, and useful pedagogical locations, it will be important to travel with a critical eye. All wild places are not equivalent. If they were, wilderness would become such a broad term that it would be useless. At the least, wilderness must be seen as a continuum—with more or less degrees of wildness.

Boreal forests, peat bogs, Scottish islands, deserts, and jungles, can all, as Cronon suggests, reflect "a more intricate web of ecological relationships." This does not mean, however, that educators must always seek field sites like these. But, at the same time, these places should not be dismissed either—lest that they become forgotten and vulnerable to rapacious development. Wild pedagogy must be clear about when, where, and what wildness we seek to nurture. Urban parks and trees in our gardens can be wondrous, but they are themselves colonized sites. And, these places of wonder can be paired with critique.[41] It is the educator's task to help people navigate these spaces and not fall into human-centric, colonial ways of being. And they must do this navigating without colonizing the place through the interpretive frameworks that humans so often bring.

In closing, we will give the last word to Gary Snyder for two reasons. First, he eloquently provides at least a partial summary of a rethought conception of wilderness. Second, he foreshadows another important element in this exploration of wild pedagogies.

> "Wild" alludes to a process of self-organization that generates systems and organisms, all of which are within the constraints of—and constitute components of—larger systems that again are wild, such as major ecosystems or the water cycle in the biosphere. Wildness can be said to be the essential nature of nature. As reflected in consciousness, it can be seen as a kind of open awareness—full of imagination but also a source of alert survival intelligence. The workings of the human mind at its very richest reflect this self-organizing wildness. So language does not impose order on a chaotic universe, but reflects its own wildness back.[42]

Finding ourselves at home in wild places—reverberating with wildness—will depend on how we carry ourselves in the world. It will rest on the language and metaphors we use, the humility we exercise when listening to more-than-humans, and the practices and pedagogies we engage in. Finding ourselves at home will depend on the etiquette we lead with.[43] This chapter has sought to trouble the ways humans impose order on the natural world. Going forward, we will continue to examine what it would mean to listen better, and what would it mean to hear wilderness reflecting its own order back upon us.

Acknowledgements *Crex crex* is the taxonomical name given to the Corncrake. We have chosen this bird to represent our collective because it was an important collaborator in this project and because its onomatopoeic name beautifully mirrors its call—a raspy crex crex. For some reason, it chooses to fly over England

and breeds in Scotland and Ireland. Presumably this is due to loss of habitat in modern England, but perhaps these birds sense some epicenter of empire there? Who is to know?

Notes

1. See, for example, Clive Hamilton, *Requiem for a Species: Why We Resist the Truth About Climate Change* (Washington: Earthscan, 2010); Elizabeth Kolbert, *The Sixth Extinction: An Unnatural History* (New York: Henry Holt and Company, 2014); John Foster, *After Sustainability: Denial, Hope Retrieval* (New York: Routledge, 2015).
2. See, for example: Naomi Klein, *This Changes Everything: Capitalism Vs. the Climate* (Toronto: Alfred A. Knopf Canada, 2014).
3. See for example: Gilles Deleuze and Felix Guattari, *What is Philosophy?* Trans. G. Burchell & H. Tomlinson (London: Verso, 1994).
4. William Cronon, "The Trouble with Wilderness: Or, Getting Back to the Wrong Nature," in *Uncommon Ground: Rethinking the Human Place in Nature*, ed. W. Cronon, 69–90 (New York: W.W. Norton, 1996): 69.
5. See for example: Emma Morris, *The Rambunctious Garden: Saving Nature in a Post-Wild World* (New York: Bloomsbury Publishing, 2013).
6. Cronon, *The Trouble with Wilderness*, 70.
7. Ibid., 69.
8. Dave Foreman, *The Great Conservation Divide: Conservation Vs. Resourcism on America's Public Lands* (Durango, CO: Ravens Eye Press, 2014).
9. James Hunter, *On the Other Side of Sorrow: Nature and People in the Scottish Highlands* (Edinburgh: Birlinn Limited, 1995): 24.
10. Ibid., 25–26.
11. Ibid., 135.
12. Reported in Hunter, *On the Other Side*, 20.
13. Note: there are twenty-six tribes have ancestral connections to Yellowstone.
14. J. Peepre, "Protected Areas and Wilderness in the North," in *Northern Protected Areas and Wilderness*, ed. J. Peepre and Bob Jickling, 5–19 (Whitehorse: Canadian Parks and Wilderness Society and Yukon College, 1994): 11.
15. Norma Kassi, "In a Panel Discussion: What Is a Good Way to Teach Children and Young Adults to Respect the Land?" In *A Colloquium on

Environment, Ethics, and Education, ed. B. Jickling, 32–48 (Whitehorse: Yukon College, 1996): 42.
16. Norma Kassi, "Science, Ethics and Wildlife Management," in *Northern Protected Areas and Wilderness*, ed. J. Peepre and Bob Jickling, 212–216 (Whitehorse: Canadian Parks and Wilderness Society and Yukon College, 1994): 215.
17. Hunter, *On the Other Side*, 47.
18. Ibid., 43–44.
19. The term "wild nature" is used here to describe the non-human-controlled world. This connects us to the old Scandinavian dichotomy between "*kulturmark*" and "*vildmark.*" "*Kulturmark*" describes the fenced "cultural land" of the agricultural pre-modern farmers. This land surrounded early agricultural dwellings, and was controlled by humans. In this fenced *kulturmark*, vegetables and crops were grown, and livestock was kept during nights. The opposite, outside the fenced and controlled area, was the waste "*vildmark.*" This wild land was seen as wild and dangerous. This is where beasts, trolls, and the mystic and mythological "spirits and entities" dwelled. *Kulturmark* was a safe place, while *vildmark* were dangerous places that required experience and knowledge to travel and survive in. However, this agricultural view of the land was not shared by the nomadic Sami people from Sápmi land in northern Scandinavia. These Sami people lived *in* the land, without this dichotomy between wild and cultivated spaces. They followed the migration of reindeer from the summer habitats the mountains to winter habitats in the lower boreal forests near the coast. For the Sami people, there was no "wilderness." There was just the free land to travel and live in. This pre-modern agricultural relationship to the land and nature has dominated our present cultural understanding of nature, while ignoring the understanding of the land by the Indigenous non-agricultural people. For the Sami people, calling the Laponia Wild Heritage Area, or Sarek National Park in Northern Sweden, the "last wilderness of Europe" is an insult to their lifestyle. Labeling their home "wilderness" is a colonial offence to their heritage. For the Sami people this "wilderness" is not a wild land, but rather a free land. Here the reindeer can roam freely and prosper. This stands in stark contrast to the more recently "cultivated" land in the coastal areas of northern Scandinavia. Here cities and villages, roads, mining, industries, agriculture, power stations and flooded land, and other modern constructions, restrict the free roaming of the reindeer, and the Sami culture. A better way to describe wilderness would be "free

nature." But, the world "free" has its own limitations in biological and ecological senses. So I will use "wild nature," despite its drawbacks.

20. Marie Battiste, "You Can't Be the Global Doctor If You're the Colonial Disease," in *Teaching as Activism: Equity Meets Environmentalism*, ed. Peggy Tripp and Linda Muzzin, 121–133 (Kingston: McGill-Queen's University Press, 2005): 124.

21. There are many analyses of dualisms. These examples are drawn from: David Abram, *The Spell of the Sensuous: Perception and Language in a More-Than-Human World* (New York: Vintage, 1997): Neil Evernden, *The Natural Alien: Humankind and Environment* (Toronto: University of Toronto Press, 1993).

22. See, for example: Rebecca Martusewicz, Jeff Edmundson, and John Lupinacci, *Ecojustice Education: Toward Diverse, Democratic, and Sustainable Communities* (London: Routledge, 2011).

23. Chet Bowers, *Education, Cultural Myths and the Ecological Crisis: Toward Deep Changes* (Albany, NY: State University of New York Press, 1993).

24. William Cronon, "Introduction: In Search of Nature," in *Uncommon Ground: Rethinking the Human Place in Nature*, ed. William Cronon, 23–56 (New York: W.W. Norton, 1996): 34.

25. Cronon, *The Trouble with Wilderness*, 88.

26. William Cronon, "Forward," in *Uncommon Ground: Rethinking the Human Place in Nature*, ed. William Cronon, 19–22 (New York: W.W. Norton, 1996): 21–22.

27. Abram, *The Spell of the Sensuous*, 1997.

28. Bill McKibben, *The End of Nature* (New York: Random House Publishing, 1989).

29. Timothy Morton, *The Ecological Thought* (Cambridge: Harvard University Press, 2012).

30. Winona LaDuke, *All Our Relations: Native Struggles for Land and Life* (Boston: South End Press, 2008).

31. Val Plumwood, "Paths Beyond Human-Centeredness," in *An Invitation to Environmental Philosophy*, ed. Anthony Weston, 69–105 (New York: Oxford University Press, 1999).

32. Plumwood, "Paths Beyond Human-Centeredness," 69.

33. Ibid., 77.

34. Sean Blenkinsop, and Laura Piersol, "Listening to the Literal: Orientations: Towards How Nature Communicates," *Phenomenology & Practice* 7, no. 2 (2013): 41–60.

35. Abram, *Spell of the Sensuous*, 1997.

36. David Grunewald, "Foundations of Place: A Multidisciplinary Framework for Place-Conscious Education," *American Educational Research Journal* 40, no. 3 (2003): 619–654.
37. Plumwood, "Paths Beyond Human-Centeredness," 101.
38. Cronon, *The Trouble with Wilderness*, 88.
39. Arne Naess with Bob Jickling, "Deep Ecology and Education: A Conversation with Arne Næss," *Canadian Journal of Environmental Education* 5, no. 1 (2000): 48–62.
40. Anthony Weston, "What If Teaching Went Wild," *Canadian Journal of Environmental Education* 9, no. 1 (2004): 11–30.
41. Michael Derby, Laura Piersol, and Sean Blenkinsop, "Refusing to Settle for Pigeons and Parks: Urban Environmental Education in the Age of Neoliberalism," *Environmental Education Research* 21, no. 3 (2015): 378–389.
42. Gary Snyder, "Language Goes Two Ways," in *The Alphabet of Trees: A Guide to Nature Writing*, ed. Christian McEwen & Mark Statman, 1–5 (New York: Oxford University Press, 2000): 1.
43. Jim Cheney, and Anthony Weston, "Environmental Ethics as Environmental Etiquette: Toward an Ethics-Based Epistemology," *Environmental Ethics* 21, no. 2 (1999): 115–134.

References

Abram, D. *The Spell of the Sensuous: Perception and Language in a More-Than-Human World*. New York: Vintage, 1997.

Battiste, M. "You Can't Be the Global Doctor If You're the Colonial Disease." In *Teaching as Activism: Equity Meets Environmentalism*, ed. P. Tripp and L. Muzzin. Kingston: McGill-Queen's University Press, 2005: 121–133.

Blenkinsop, S., and L. Piersol. "Listening to the Literal: Orientations: Towards How Nature Communicates." *Phenomenology & Practice* 7, no. 2 (2013): 41–60.

Bowers, C. *Education, Cultural Myths and the Ecological Crisis: Toward Deep Changes*. Albany: State University of New York Press, 1993.

Cheney, J., and A. Weston. "Environmental Ethics as Environmental Etiquette: Toward an Ethics-Based Epistemology." *Environmental Ethics* 21, no. 2 (1999): 115–134.

Cronon, W. "Introduction: In Search of Nature." In *Uncommon Ground: Rethinking the Human Place in Nature*, ed. W. Cronon. New York: W.W. Norton, 1996a: 23–56.

———. "The Trouble with Wilderness: Or, Getting Back to the Wrong Nature." In *Uncommon Ground: Rethinking the Human Place in Nature*, ed. W. Cronon. New York: W.W. Norton, 1996b: 69–90.

Deleuze, G., and F. Guattari. *What Is Philosophy?* Translated by G. Burchell & H. Tomlinson. London: Verso, 1994.

Derby, M., L. Piersol, and S. Blenkinsop. "Refusing to Settle for Pigeons and Parks: Urban Environmental Education in the Age of Neoliberalism." *Environmental Education Research* 21, no. 3 (2015): 378–389.

Evernden, N. *The Natural Alien: Humankind and Environment*. Toronto: University of Toronto Press, 1993.

Foreman, D. *The Great Conservation Divide: Conservation Vs. Resourcism on America's Public Lands*. Durango, CO: Ravens Eye Press, 2014.

Foster, J. *After Sustainability: Denial, Hope Retrieval*. New York: Routledge, 2015.

Grunewald, D. "Foundations of Place: A Multidisciplinary Framework for Place-Conscious Education." *American Educational Research Journal* 40, no. 3 (2003): 619–654.

Hamilton, C. *Requiem for a Species: Why We Resist the Truth About Climate Change*. Washington: Earthscan, 2010.

Hunter, J. *On the Other Side of Sorrow: Nature and People in the Scottish Highlands*. Edinburgh: Birlinn Limited, 1995.

Kassi, N. "Science, Ethics and Wildlife Management." In *Northern Protected Areas and Wilderness*, ed. J. Peepre and Bob Jickling. Whitehorse: Canadian Parks and Wilderness Society and Yukon College, 1994: 212–216.

———. "In a Panel Discussion: What Is a Good Way to Teach Children and Young Adults to Respect the Land?" In *A Colloquium on Environment, Ethics, and Education*, ed. B. Jickling. Whitehorse: Yukon College, 1996: 32–48.

Klein, N. *This Changes Everything: Capitalism Vs. the Climate*. Toronto: Alfred A. Knopf Canada, 2014.

Kolbert, E. *The Sixth Extinction: An Unnatural History*. New York: Henry Holt and Company, 2014.

LaDuke, W. *All Our Relations: Native Struggles for Land and Life*. Boston: South End Press, 2008.

Martusewicz, R., J. Edmundson, and J. Lupinacci. *Ecojustice Education: Toward Diverse, Democratic, and Sustainable Communities*. London: Routledge, 2011.

McKibben, B. *The End of Nature*. New York: Random House Publishing, 1989.

Morris, E. *The Rambunctious Garden: Saving Nature in a Post-Wild World*. New York. Bloomsbury Publishing, 2013.

Morton, T. *The Ecological Thought*. Cambridge: Harvard University Press, 2012.

Næss, A. with B. Jickling. "Deep Ecology and Education: A Conversation with Arne Næss." *Canadian Journal of Environmental Education* 5, no. 1 (2000): 48–62.

Peepre, J. "Protected Areas and Wilderness in the North." In *Northern Protected Areas and Wilderness*, ed. J. Peepre and B. Jickling. Whitehorse: Canadian Parks and Wilderness Society and Yukon College, 1994: 5–19.

Plumwood, V. "Paths Beyond Human-Centeredness." In *An Invitation to Environmental Philosophy*, ed. A. Weston. New York: Oxford University Press, 1999: 69–105.

Snyder, G. "Language Goes Two Ways." In *The Alphabet of Trees: A Guide to Nature Writing*, ed. Christian McEwen and Mark Statman. New York: Oxford University Press, 2000: 1–5.

Weston, A. "What If Teaching Went Wild." *Canadian Journal of Environmental Education* 9, no. 1 (2004): 11–30.

3

On the Anthropocene

The Crex Crex Collective

Abstract This chapter focuses on three key points. First, the world has changed, with destructive consequences for many, will continue to change, and will not return to situation "normal." That is, it will not return to global temperatures or species abundance and fluctuations that fall within the kinds of background levels experienced by generations of humans. This terrifying transformation has been labelled "The Anthropocene." While it is acknowledged that this term is contentious it is used here for its capacity to do useful work. Second, any educational

---

The Crex Crex Collective includes: Hebrides, I., Independent Scholar; Ramsey Affifi, University of Edinburgh; Sean Blenkinsop, Simon Fraser University; Hans Gelter, Guide Natura & Luleå, University of Technology; Douglas Gilbert, Trees for Life; Joyce Gilbert, Trees for Life; Ruth Irwin, Independent Scholar; Aage Jensen, Nord University; Bob Jickling, Lakehead University; Polly Knowlton Cockett, University of Calgary; Marcus Morse, La Trobe University; Michael De Danann Sitka-Sage, Simon Fraser University; Stephen Sterling, University of Plymouth; Nora Timmerman, Northern Arizona University; and Andrea Welz, Sault College.

Bob Jickling (bob.jickling@lakeheadu.ca) is the corresponding author.

B. Jickling (✉)
Lakehead University, Thunder Bay, ON, Canada
e-mail: bob.jickling@lakeheadu.ca

© The Author(s) 2018
B. Jickling et al. (eds.), *Wild Pedagogies*, Palgrave Studies in Educational Futures,
https://doi.org/10.1007/978-3-319-90176-3_3

conception and delivery that results in inculcation into dominant cultural norms will do nothing to change the current trajectory nor prepare learners for the new reality. Finally, no one really knows how to move forward in the best possible way. This isn't meant to sound despairing, rather it signifies, that we're in a time calling for bold experimentation and imagination.

Keywords Anthropocene • Education • Environmental • Geostory • More-than-human

The earth is changing rapidly. Atmospheric carbon dioxide has now exceeded 400 parts per million, and continues to rise. At present, there is no realistic strategy in place to make the reductions necessary to avoid what most climate scientists consider "catastrophic" climate change. Species loss has been equally dramatic. Some reports, such as a recent publication in the prestigious journal *Science*,[1] suggest that current extinction rates are as much as 1000 times greater than background rates. These extinction rates are human-caused, as are the current dramatic increases in Earth's average temperatures. These observations are even more disturbing when considering Bruno Latour's bleak observation that the real drama is behind us—that we have already crossed planetary boundaries that some scientists have identified as ultimate barriers not to be overstepped.[2]

These planetary boundaries that "must" not be crossed, have not been established lightly. Science is typically a conservative enterprise; it actively seeks to avoid false positives and alarmist rhetoric. Some hard-nosed palaeontologists, for example, tell us that current rates of species loss do not yet qualify as mass extinctions.[3] But, ominously, they *are* prepared to predict that loss of all species that are now considered "critically endangered" *would* propel the world into a state of mass extinction. Should that happen, it would comprise a paleontological event of epochal magnitude. Given that there appears to be no abating of species loss in sight, other scientists are willing to argue that we—and that means all beings on Earth—are in fact living in a new geological epoch: the Anthropocene.[4] At this time, it seems that humans, pigeons, and crows are expanding their ranges, and not much else is doing well.[5] Nearly every ecological system is in decline, and scientists also expect rapidly self-exacerbating feedback loops to unravel much of what remains. And these processes remain largely unpredictable. As French philosopher Michel Serres has put it, "…the Earth is quaking… because it is being transformed by our doing."[6] He goes on to say:

> it depends so much on us that it is shaking and that we too are worried by this deviation from expected equilibriums. We are disturbing the Earth and making it quake! Now it has a subject once again.[7]

The idea the Earth "has a subject," again, is important to our concerns. It is our sense that Serres is suggesting that Earth, in its quaking, has jutted through the idea of objectivity and made its active presence known. We will return to this discussion shortly. In the meantime, there are educational questions about how this epoch should be named, discussed, approached, and addressed.

Does labelling this terrifying transformation "The Anthropocene" do the educational work we need? Maybe. We think it can, but this will require care. As this term becomes more present in every day conversation—and in cultural artefacts—it becomes normalized and, over time, it could lose its disruptive and generative possibilities. At face value, inserting "Anthropocene" into the cultural milieu might serve to shake people into action by highlighting the severity of the calamity. We suppose this is possible, but invoking ecological crisis—since the dawn of environmental education, and before—does not seem to have had much impact, pedagogically or otherwise. This evocation does not seem to offer sufficient traction to disrupt traditional pedagogical instincts nor educational theories.

In the end, we have chosen to use Anthropocene because we believe it has capacity to do some useful work. It seems that the emergence of a new paleontological era must be accompanied by a new story about Earth—a new geostory. And, this will be a story that is told by myriad tellers. Educationally, it will be important to examine how to be better tellers and listeners. It will also be important to understand how this emerging geostory can contribute to new geopolitical understanding. To

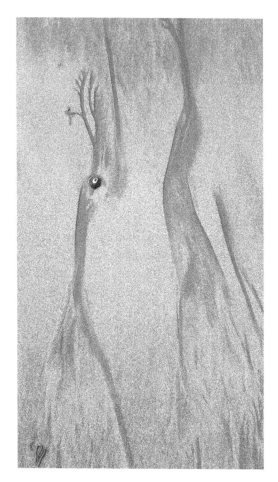

Image 3.1 Sands of deep time. Photo credit: Hansi Gelter

pick up on the idea that there are, indeed, multiple authors in a new geostory, we turn to another French philosopher, anthropologist and sociologist, Bruno Latour (Image 3.1).

Latour picks up on Serres and puts into words what we humans see—especially at a time when wildfires are scorching Earth and cyclones are flooding her. He says, "Earth has become—has become again!—an active, local, limited, sensitive, fragile quaking, and easily tickled envelope."[8] Here, Latour acknowledges that Earth has always been an active presence, but in a role that has often been overlooked and denied in modernist thinking, and culture. Today, Earth's agency is visible, often dramatic, and it can no longer be ignored. For Latour and others, she has reclaimed the character of a full-fledged actor—an agent of history. In employing the term Anthropocene, he sees Earth as participating in writing the script of our common *geostory*[9]—a narrative that, try as they might, humans can no longer write alone. The problem is how those in philosophy, science, politics, and literature can share space and tell such a story.

For those of us writing this book, our human part in telling this geostory is also profoundly educational. What does it mean, educationally, to participate in telling a story where we share agency with Earth herself? What does it mean, pedagogically, when we, the human teachers, are not in total control of the script? What could it mean at the dawn of the Anthropocene to bring wilderness and education together in concept and practice? And, how does education support and challenge learners who want to challenge modernist assumptions about control, and human as the elite species?

To begin with, there are a couple of considerations. The first involves how writing a geostory can challenge how we see, feel, and talk about the world, and the assumptions those historical perceptions have long rested upon. The second challenges us to develop and engage with a sensory awareness that can help us to better understand our position in the world.

For people to now say that Earth has a subject is to upend a worldview that has dominated human relationships with her since the Scientific Revolution. It confuses categories once thought of as subjects and objects where, in this scheme, Earth has largely been de-animated. It has been reduced to a backdrop upon which humanity acts. As Latour points out, this is the frightening meaning of "global warming." Here he argues,

> human societies have resigned themselves to playing the role of the dumb object, while nature has unexpectedly taken on that of the active subject! …it is *human* history that has become frozen and *natural* history that is taking on a frenetic pace.[10]

In considering this example, it is clear that the hubristic dream of human control and oversight can no longer be sustained. In important ways, humans have never actually been in control. And, Earth, demonstrably, has always had subject-hood and agency in shaping geostories.

If humans are implicit actors, largely destructive and in denial of the consequences, what else might be required in taking our place alongside other agents and collaborators in the writing of our geostory? How can we be more attentive to their agency? For, as Latour assures us, "*As long as they act, agents have meaning.*"[11] But, not all meaning arises in human terms. And, not all story telling arises through human language. A more expansive and inclusive story can arise as a consequence of being fully present in an articulated and active world. For Latour, it will be impossible to tell our common geostory without everyone—including educators, learners, and more-than-humans—having the space to share their own perspectives and being heard in their own ways. For us this suggests that essential learning will require breaking down those boundaries that have helped us believe that we were not in the world and discovering how to listen well to other-than-humans—it will require being in the world, and being with the world. Becoming Earthbound.

In some senses this book and the colloquium leading to it represent early attempts to enact some of the challenges outlined here. The colloquium immersed itself in a place. It was hosted on a boat that travelled through a geographical and cultural place. Much work was conducted out of doors. Natural and cultural historians were present to enrich our collective experiences in this landscape. Writing daily manifest entries constituted conscious acts of listening to more-than-human subjects—and being present. Interruptions to human discourses were observed, noted, sought. And, written anecdotes encouraged recounting of specific experiences in particular places. These have been our experiments. Still, being present in the place and conceptualizing wild pedagogies were not always compatible. There was a lot going on; there remains a lot to do.

Fulmar Petrels

Turns out the soft grey birds on the grassy cliffs of Staffa are fulmars. Though they look like gulls they are actually petrels—or sometime fulmar petrels. Snubby beaks and cute heads—not as sleek as a gull, with wide nostrils that exude salt, Doug says, when they are out at sea. They are klutzy on land. One saw me peering over the cliff and got nervous so she swung away from her perfect nest site and swooped out across the small bay, over to the crenulated cliffs on the other side of the creek. I thought she'd pause but instead she swooped back around and tried to land next to the nest of her neighbour. They didn't seem to say anything, or at least I never heard it but subsequently the neighbour peered up too, with large gentle eyes keeping a watch on me, lying above on plush grass and primroses. Seaweed smells rose up the cliffs, decomposing in the sun. The displaced fulmar flapped and flapped, trying to gain her footing but she couldn't find her balance on the almost vertical grassy cliff. Wings outstretched, she waivered and finally gave up and flew off again, sweeping out over the water. She floated up the updraft on the other cliff and beat her wings a couple of times as she swooped back swinging around close to her friend and then away and back out again. Around one more time before landing more securely a tad further over on the wall. They must have conferred on potential landing spots. I couldn't hear them. Larks sang, a tourist boat chugged far too loudly. Thrift and deep grass cushioned my body. The fulmar flew off again. She seemed more relaxed.

I watched her and changed focus to the seaweed swinging, feathery, in the tide. A shag flew in, low and straight. The creek trickled. The tourists checked out a bird colony on a nearby rock and chugged away. My fulmar friend alighted on her own nesting spot. She tripped, webbed feet clumsy on grass and rock. She's found a good place. Once she lurches past a little rock, she's got a flat grassy nook. No eggs yet, but it will be perfect. She checks me out again but I can tell she's no longer worried. How do I know that? No idea. Maybe mirror neurons. But when I stopped staring at them and allowed my gaze to wander over the bay, past the shape of the cliffs and the tide and out to the rocky islands, the boats, inland up the creek... in that relaxed sweep and context, the birds decided I wasn't hunting. I went from intruder alert to visitor alert. Not exactly background noise like the tourist boat. Or maybe like a seal; a mammal cohabiting space, but not on either end of the food chain spectrum.

It was so relaxing in the sun and grass, with my company, I didn't really want to get up again. But the boat's leaving in 10 minutes or so, time shifted from the seasonal pace of early spring fulmars to the pressing regime of a journey.

In the forgoing, we explained why we use the term "Anthropocene" at this time. We certainly are not attempting to reify the existence of this epoch, or the term itself. We use it to do some useful work. Emerging evidence clearly indicates that humans, particularly the modernist versions, cannot control the Anthropocene's developing geostory; they cannot co-opt its writing, and they are not the only tellers. Indeed, the telling will require renegotiation of how stories are told and who constitutes a teller.

While we believe that framing our times as the Anthropocene can do work for us, it is important to recognize that this term is also problematic. This becomes a pressing matter as major international geological societies are normalizing the term and as it enters the realm of everyday conversation. For it to continue to be useful—and not descend into cliché, just another burden, or worse—connecting the Anthropocene to a new geostory will need to be an on-going process.

In spite of interpretations such as that offered by Latour, some critics worry that the Anthropocene still places humans at the centre of conversations. For them, human *hubris* could lead to we-broke-it-but-we-can-fix-it, or see-how-special-we-are-we-changed-the-world, attitudes. Ultimately, according to these views, Anthropocene-talk will not be sufficient to dislodge human-centredness run amok. In some measure, this is probably true. However, there is always a likelihood that moves will be made to co-opt whatever term is used to describe our present era. And while there will always be critics, sceptics, and deniers unwilling to cede control, challenging human-centredness will require vigilance, and will be an on-going task.

Critics also worry that normalizing the Anthropocene might mask particular economic, technological, cultural, and material realities that gave rise to the current globalizing culture. In fact, there is a large interlocking network of causal factors. To label a problem in a singular way overshadows a more *ecological* conception of life. In response some scholars have suggested the current epoch be labelled the Capitalocene or the Chthulucene (from the Greek *chthon*, meaning Earth). For Donna Haraway, the Anthropocene just does not tell a nuanced enough story. It makes opaque the particular roles of global capital, or colonial orientations, or the patriarchy in the environmental crisis. Thus, it allows the more egregious perpetrators to avoid direct responsibility as they slide into the generic position of all humans. In a move that both affirms and troubles the concept of the Anthropocene, she would like to see a thousand names of something else to erupt out of present use of this term.[12] And she adds a critically important point. This issue is not just about naming it is also about imagining, developing, and doing new kinds of work—or in her terms, labour—which in turn can be used to envision and create new conceptions of, and relationships with, nature.

With these points in mind we acknowledge that the Anthropocene is contentious. Conversations about how humans see themselves in the world are erupting and we need to participate in renegotiating the new geostory of our time. And, we are just beginning to understand that we have co-authors and co-tellers. Terminology will develop and change over time. In the meantime, we agree with Haraway that it is important to do new kinds of work, and in our case this involves new kinds of teaching and learning, to enable our participation in these re-negotiations. However, this is a dynamic conversation and out of each change in the present work we may find evermore-effective and ecological approaches to continue our labour. We encourage readers to keep an eye on developments in this conversation.

Whether we are actually in a new geological era called the Anthropocene, or just on the brink of it, seems moot. We raise these possibilities, however, for two reasons. First, the world has changed in destructive ways, will continue to change, and will not return to situation "normal." That is, we will not return to global temperatures or species abundance and fluctuations that fall within the kinds of background levels experienced by generations of humans just a short time ago. Second, given this change, any educational conception and delivery that results in inculcation into present cultural norms, or slipping and sliding around these norms, will do nothing to change the current trajectory nor prepare learners for the new reality. A critical task will be to see where this recapitulation of present norms might be happening. This returns us to the old trope, especially important for educators to ponder; we cannot "solve problems" by using the same kind of thinking that created the "problems" in the first place. And we will add, we cannot solve these problems by "*being*" the same people that created the problem.

Finally, no one really knows how to move forward in the best possible way. We do not mean for this to sound despairing, rather it signifies, to us, that we are in a time calling for bold experimentation and imagination.[13] To be sure, we are not talking about laboratories or the scientific model of experimentation. Rather we encourage a more general interpretation—where we let teachers and learners try things out. This will require giving them the social, psychological, and phenomenological room that they need to explore—and renegotiate—new ideas. It will also require the conceptual, experiential, and physical freedom to move and think. With these preconditions in place, we can have considerably more wild pedagogy. Here individuals and groups *can* actually begin to participate in new practices and new relationships through everyday practices.[14]

What we are calling for will be creative, courageous, and radical—because this is what our times require. But this does not mean that we are proposing an anything goes free-for-all. The kinds of educational experiences required will need to imagine new relationships with nature, to take into account the agency of the more-than-human, to be flexible and able to change as new thinking makes new ideas and possibilities apparent, and they will need to be carefully planned and mentored. Our ideas for how to facilitate these educational experiences are discussed further in the next chapter *On Education*, and in the following chapter, *Six Touchstones for a Wild Pedagogy*.

Acknowledgements *Crex crex* is the taxonomical name given to the Corncrake. We have chosen this bird to represent our collective because it was an important collaborator in this project and because its onomatopoeic name beautifully mirrors its call—a raspy crex crex. For some reason, it chooses to fly over England and breeds in Scotland and Ireland. Presumably this is due to loss of habitat in modern England, but perhaps these birds sense some epicenter of empire there? Who is to know?

Notes

1. Stuart Pimm et al., "The Biodiversity of Species and Their Rates of Extinction, Distribution, and Protection," *Science* 344 (2014): 6187.

2. Bruno Latour, "Agency at the Time of the Anthropocene," *New Literary History* 45, no. 1 (2014): 1–18.
3. A. Barnosky et al., "Has the Earth's Sixth Mass Extinction Already Arrived?" *Nature* 471 (2011): 51–57.
4. See, for example: Editorial, "The Human Epoch," *Nature* 473 (2011): 254: Paul J. Crutzen, "Geology of Mankind," *Nature* 415 (2002): 23.
5. Sean Blenkinsop, and Laura Piersol, "Listening to the Literal: Orientations: Towards How Nature Communicates," *Phenomenology & Practice* 7, no. 2 (2013): 41–60: Michael Derby, Laura Piersol, and Sean Blenkinsop, "Refusing to Settle for Pigeons and Parks: Urban Environmental Education in the Age of Neoliberalism," *Environmental Education Research* 21, no. 3 (2015): 378–389.
6. Michel Serres, *The Natural Contract*, trans. Elizabeth MacArthur and Robert Paulson (Ann Arbor: University of Michigan Press, 1995): 86
7. Ibid., 86.
8. Latour, "Agency at the Time of the Anthropocene," 4.
9. Ibid., 3.
10. Ibid., 13.
11. Ibid., 14.
12. Donna Haraway, *Staying with the Trouble* (Durham: Duke University Press, 2016).
13. Bob Jickling, "Normalizing Catastrophe: An Educational Response," *Environmental Education Research* 19, no. 2 (2013): 161–176.
14. This experimental vision is derived from: Anthony Weston, "Before Environmental Ethics," *Environmental Ethics* 14, no. 4 (1992): 321–338.

References

Barnosky, A., N. Matzke, S. Tomiya, G.O. Wogan, B. Swartz, T.B. Quental, C. Marshall, et al. "Has the Earth's Sixth Mass Extinction Already Arrived?" *Nature* 471 (2011): 51–57.
Blenkinsop, S., and L. Piersol. "Listening to the Literal: Orientations: Towards How Nature Communicates." *Phenomenology & Practice* 7, no. 2 (2013): 41–60.
Crutzen, P.J. "Geology of Mankind." *Nature* 415 (2002): 23.

Derby, M., L. Piersol, and S. Blenkinsop. "Refusing to Settle for Pigeons and Parks: Urban Environmental Education in the Age of Neoliberalism." *Environmental Education Research* 21, no. 3 (2015): 378–389.

Editorial. "The Human Epoch." *Nature* 473 (2011): 254.

Haraway, D. *Staying with the Trouble*. Durham: Duke University Press, 2016.

Jickling, B. "Normalizing Catastrophe: An Educational Response." *Environmental Education Research* 19, no. 2 (2013): 161–176.

Latour, B. "Agency at the Time of the Anthropocene." *New Literary History* 45, no. 1 (2014): 1–18.

Pimm, S.L., C.N. Jenkins, R. Abell, T.M. Brooks, J.L. Gittleman, L.N. Joppa, P.H. Raven, C.M. Roberts, J.O. Sexton. "The Biodiversity of Species and Their Rates of Extinction, Distribution, and Protection." *Science* 344 (2014): 6187.

Serres, M. *The Natural Contract*. Translated by E. MacArthur and R. Paulson. Ann Arbor: University of Michigan Press, 1995.

Weston, A. "Before Environmental Ethics." *Environmental Ethics* 14, no. 4 (1992): 321–338.

4

On Education

The Crex Crex Collective

Abstract This chapter presents educators with a conundrum: *how to change educational systems so that they can in turn promote learning relevant to and commensurate with the multiple crises we face, without being co-opted by dominant cultural norms.* Instead of seeking to integrate environmental and sustainability education into existing educational institutions, the challenge is rather the reverse. The task at hand is really to renegotiate, in conjunction with Earth and the more-than-human world, the idea and practice of education itself. Beneath what appear as crises, such as climate

---

The Crex Crex Collective includes: Hebrides, I., Independent Scholar; Ramsey Affifi, University of Edinburgh; Sean Blenkinsop, Simon Fraser University; Hans Gelter, Guide Natura & Luleå, University of Technology; Douglas Gilbert, Trees for Life; Joyce Gilbert, Trees for Life; Ruth Irwin, Independent Scholar; Aage Jensen, Nord University; Bob Jickling, Lakehead University; Polly Knowlton Cockett, University of Calgary; Marcus Morse, La Trobe University; Michael De Danann Sitka-Sage, Simon Fraser University; Stephen Sterling, University of Plymouth; Nora Timmerman, Northern Arizona University; and Andrea Welz, Sault College.

Bob Jickling (bob.jickling@lakeheadu.ca) is the corresponding author.

B. Jickling (✉)
Lakehead University, Thunder Bay, ON, Canada
e-mail: bob.jickling@lakeheadu.ca

© The Author(s) 2018
B. Jickling et al. (eds.), *Wild Pedagogies*, Palgrave Studies in Educational Futures, https://doi.org/10.1007/978-3-319-90176-3_4

change and species extinctions, a more profound crisis lies in the way that many humans relate to the world—that is the dominant modernist way of being in the world. A renegotiated and renewed vision of education must include structures, curricula, and pedagogies that are fundamentally disruptive to these ways of being.

Keywords Curriculum • Education • Environmental • Pedagogies • Touchstones • Wild

Many people who have been initially drawn to wild pedagogies share similar experiences. In their educational lives, they know that the most significant learning—learning that has actually been transformative in the way that their lives have been lived—has often been encountered outside of formal education or at the margins of their schooling by brave, insightful, and rebel teachers. Sometimes this learning has even occurred without a teacher, in a conventional sense. By rebel teachers we mean those often quiet educators that work alongside many others in response to the ecological and social injustices, even catastrophes, of our era. They know that they must find something to do in response to these challenges. And they are the ones that often generate the experiences that live on in learners' lives.[1]

We suspect that those drawn to wild pedagogies, know that bringing different visions of education—or ideas like "nature as co-teacher"—into the mainstream is easier said than done. Many rebel teachers and education students know this well, particularly those who come to education with real-world experiences outside of the kind teaching that most often occurs in formal settings. Some have been outdoor and environmental educators who worked in camps, for wilderness operators, with non-governmental organizations, and as interpreters. Some have been involved in social justice issues and have worked with grassroots organizations committed to fairness, justice, and equity. Some have worked in alternative programmes both within and in response to mainstream education. Some have imaginatively toiled within the system itself, closing their doors (or maybe opening them to the playgrounds and parks beyond) and doing their own thing.

What unites these teachers is a passion for making a difference. Their experiences have taught them that education can be more inspirational, and pedagogy more transformational, than that which what they may have experienced during their own schooling. Yet students and teachers often struggle with an education system that has pushed to the side much of what they most value. As it turns out, most of their best learning experiences do not fit neatly into the prescribed "teachable" subjects.

Perhaps the keyword here is "prescribed." When educational experiences are prescribed student learning tends to serve the ends of an education process based on predetermined outcomes—preferably those that are measurable. There is plenty of research suggesting that curriculum content and pedagogical strategies are bent to align with testable outcomes. Learning that is less amenable to testing is edged out.[2] Even in education faculties, enormous efforts are made to prescribe and control narrow versions of how education is defined and how it looks. Yet, as Arjen Wals reminded us, "What you can't measure still exists." And despite curriculum control and testing pressures, many committed teachers find ways to resist, to create, and wiggle into spaces for what they consider "real teaching." For the immeasurable.[3]

One such space for resisting control is in curriculum theorist Elliot Eisner's[4] "expressive outcomes." These are the consequences of activities that are planned to provide rich learning opportunities, but without explicit or precise objectives. The aim here is to shift emphasis away from evaluation and back to considering what good learning opportunities would look like—first and foremost. As Eisner says, "The tack taken with respect to the generation of expressive outcomes is to engage in activities that are sufficiently rich to allow for a wide, productive range of educationally valuable outcomes."[5] A key point is that focus shifts to creating good learning environments where outcomes arise from a combination of activities and context such that educators cannot possibly pre-determine everything that will happen and what learners might take away. They are in important ways self-willed, uncontrolled, and even wild.

On a cautionary note, this wild education is not the same as "going wild" or "out of control" in the more colloquial sense. The activities are carefully selected—and in this sense remain mediated. Yet they still give learners' innate wildness and curiosity more space to flourish. Intriguingly, this kind of pedagogy can inspire the opposite of the out of control wildness as learners find ways to exercise their own self-regulation. For, with freedom like this comes responsibility to oneself, to others, and to the learning project itself.

A Conundrum

For the last forty years or so, people involved in environmental and sustainability education have been occupied with questions about how to integrate their ideas, their approaches to education, and their pedagogies into the mainstream. To some extent, this effort has been successful. Globally, there are policies, programmes, curricula, pedagogic practices, and research agendas centring on environmental and sustainability education that together form a significant movement. However, despite this achievement, the overall pattern has been one of accommodation. The dominant educational paradigm is a powerful force capable of absorbing and bending new ideas back towards the *status quo*. So, while environmental education may appear to provide contrary and critical perspectives, the culture of mainstream education can render it largely ineffectual.

This kind of pervasive force that protects the *status quo* can take differing ideas and minimize—or flatten out—the inherent contradictions. The flattening out of conflicting points of view, or controversy, or differing social assumptions—it can be framed in many ways—does not come about through direct challenges. Rather, it comes about "through their wholesale incorporation into the established order, through their reproduction and display on a massive scale."[6] Rather than confront the challenging ideas of our times they are incorporated into our cultural texts. For example, instead of confronting ideas about sustainability, neo-liberal forces have embraced the term to talk about their own interests, like sustainable growth, or sustainable mining, or even sustainable excitement. When sustainability talk becomes ubiquitous, and is used to represent such a wide variety of interests, it loses its ability to disrupt dominant attitudes and assumptions. Similarly, environmental education can be embraced within mainstream education, but only if it can be shown to help meet existing learning outcomes, or otherwise support the mainstream system. These forces don't really "want" you to make a difference in any fundamental way. They want you to conform to acceptable norms. Today we call it co-option.

In many ways, much of environmental, sustainability, and outdoor education has been conforming, rather than rebellious. The mainstream culture remains, by and large, unaffected. Indeed, we as authors and members of environmental, sustainability, and outdoor education communities, acknowledge that to varying degrees, these educational approaches have been co-opted—their potential to act as grit in the oyster has been curtailed. They have underperformed. They have been made safe.

The mainstream system adjusts sufficiently to accommodate any challenge, but in a way that ensures the system remains essentially the same. Mainstream education and the educational paradigm that underpins it—informed in the past 20–30 years by neo-liberalism—is itself a resilient system. It does not change radically, or easily.

In the meantime, we are convinced of two truths. One is that the world—the planet—is in an unprecedented climacteric state that threatens the survival of myriad species including our own. Second, courageous learning and teaching are absolutely key to the bold experimentation and generation of imaginative possibilities required for fundamental change and for supporting learners through such changes.

So, here we are presented with a conundrum: *how to credibly change educational systems so that they can in turn promote learning relevant to and commensurate with the multiple crises we face, without being co-opted?* Instead of seeking to integrate environmental and sustainability education into existing educational institutions, the challenge is rather the reverse. The task at hand is really to renegotiate, in conjunction with all-our-relations,[7] the idea and practice of education itself. Beneath what appear as crises, such as climate change and species extinctions, a more profound crisis lies in the way that many humans relate to the world—that is the dominant modernist way of being in the world. A renegotiated and renewed vision of education must include structures, curricula, and pedagogies that are fundamentally disruptive to these ways of being.

And this is where the ideas of wild pedagogies—as characterised in this book—comes into play: going beyond the norm, achieving critical reflexivity, seeing the shoreline and puffins as co-teachers, in order to facilitate different practices, new insights, novel ways of working that spur imagination and vision, nourish creativity, build connection, and counter feelings of alienation and isolation. But at heart, we affirm a role for nature as co-teacher. And we affirm the wild spirit of people, particularly young people, which are all too often impoverished by testing and admonishment around "performance"—but whose passion for life needs to be freed up and nurtured if they are to be able to secure a future.

> ### Stepping Out
>
> The young toddler, not much more than one year old, and not walking yet, is at the top of the stairs. Let's call her Amy. She's looking down, wondering about whether she wants to try to descend. Before she can gently lower one podgy limb to the step below—which is a long way down for her—mother picks her up. Too dangerous, mum thinks. She might be right.
> But supposing the mother constrained all her toddler's instincts to explore her environment, to experiment, to make sense of her world through touch, smell, hearing, trying things out, and to take risks? Of course, the baby would be impoverished.
> Years later, Amy and her peers will likely pass through an education system that constrains possibilities, that disallows surprise discoveries, that frowns on what is deemed error, that rewards those that conform to the rigidity of set schemes, learning outcomes, and assessments.
> But look! Amy wants to follow up something that has piqued her curiosity, she has questions that don't seem to be on the syllabus, she wants to be in the outdoors, and experience wonder. She is interested in ethics, in politics, in the state of the world. In the kind of world she and her friends are inheriting, and the kind of futures they might have. Well, she is told, that's all great, but it won't help you pass your exams.
> So she makes it to her 20s. She's qualified. And so are lots of her friends. But Amy feels something is missing in herself. She's unsure who she is. She is a little apprehensive when she finds herself in an unfamiliar environment. She lacks social skills. She knows she is ignorant of nature, beyond simple labels—daisy, dandelion. She feels deeply—she wants to make a positive difference in the world, but feels unprepared and ill-equipped.

Wildness of Learners

Amy's experience, described in the "stepping out" vignette, is not unusual. For her, and thousands like her, the culture, values, and practices of the mainstream educational system have become increasingly limiting in recent decades. More instrumental, more technocratic, more mechanistic, under the wave of neo-liberal thinking that has seeped into and re-shaped very area of public life, and even reconfigured what counts as normal discourse.

It was once very different. We talked of the intrinsic value of education, of child-centred learning; of holistic approaches and of head/hands/ and heart; rounded education; collaborative and active pedagogies; teachers as learners; teachers as curriculum developers and collaborators; nature in the classroom; the classroom in nature; place-based education; art, music and drama held in equal esteem as more technical subjects, whole school approaches. And so on.

This was an education that honoured its etymological origins—the Latin *educare*, meaning to rear or foster, and *educere* which means to draw out or develop. Now, educational thinking, purpose, policy and practice has been squeezed. And it has been reduced to two dimensions. Horizontally, the bandwidth of what counts as important and legitimate has narrowed. So the arts are seen as increasingly marginal in the drive towards "harder" and "more technical" subjects. The vertical dimension is the depth of the learning experience. Here, the emphasis is on first order learning—content-based learning that is examinable and assessable. So deeper learning—that changes your perception, that touches your values, your sense of self, your relationship with nature and others—is not important.

Yet, given the unprecedented and multiple crises that face societies across the globe, this is the kind of education we need to embrace. One that allows and encourages boundaries to be questioned, broken, penetrated, or hurdled over—whether relating to curricula, teaching methods, assessment, or more deeply, to perception, values, attitudes and worldviews. We desperately need to let our latent (but hitherto squashed) potential be realised. To flourish. Creative, inventive, collaborative, explorative, and—importantly—risk taking, into the unknown. Wild pedagogies. Stepping out.

The Joy of Not-Knowing

As a group of authors travelling together in Scotland, discussing possibilities for wild pedagogies, we were also experiencing new landscapes, making new friends, and enjoying each other's company. At the same time we were wrestling with concepts, writing, and trying to crystallise ideas! These are all important activities that would be part of commonly accepted forms of scholarship. But, we also sought include the natural world, as an interlocutor, and an active contributor in our work. This process was, and continues to be incomplete. However, we did find that immersion in the physical space enabled our coming to nature. And we found the entry of the other-than-human was aided by approaching the place with an open, generous, attentive, curious, and maybe even welcoming mind.

In the string of qualities of mind just listed, perhaps the idea of an "open" mind is the most contentious, confusing, and even paradoxical. For example, as authors, our heads were full of stuff—conceptual stuff, writing stuff, and memories of past experiences. And, this stuff sometimes played out in unfortunate ways. For example, one member of our own group noticed, during a walk on an island stopover, two other participants vigorously discussing wild pedagogies. Suddenly, the rest of the group stopped to listen to the howling sounds of a seal colony off in the distance. The discussants stopped, too, but continued their conversation unaware that the others were trying to listen to these voices of nature. It is easy to overwhelm nature with mind stuff—to be enclosed in a social bubble.

> ### A Welcome Interruption
>
> *As the light faded, shrouding the shoreline, my social bubble was interrupted by a distant hooting sound coming from the shore. To me it sounded just like the beginning call of a barred owl, a familiar sound from home. My social bubble partner and I walked to the stern of the ship and found others listening to the calls. Doug, who helped us all become more aware of the Scottish wild voices, shared that the calls came from a colony of seals. The sound was much clearer at the stern and I for some unexplained reason sent*

out my barred owl call. It seemed like the seals responded as their calls intensified. Whether or not they actually responded to me, I felt a connection that I cannot explain—it bordered on elation. This same feeling envelops me in the spring when the world awakens; the symphony of peepers as dusk approaches, the call of the sandhill cranes, the pink hue of the spring beauties. This listening is something I would like to nurture more in my life—both personally and in my work with early childhood education students and children.

Image 4.1 A seal presence. Photo credit: Hansi Gelter

We do not think our experience is unique. Even during educational activities set within physical landscapes it is easy to slip into this kind of bubble. Part of this bubble can be social. Travelling in wild places can be full of activity—monitoring each others' well being, finding the route, sharing food, anticipating difficulties, yarning about last time. These are all important; these are not comfortable places to make a mistake. Beyond these tasks, socialisation often involves other cultural activities—singing favourite songs, talking about politics and pop culture, and joking around. These things are important, too, but they are not enough. And, paradoxically, they become themselves bubbles that filter out the voices of nature. They sometimes overwhelm the intended purposes of journey even when the point is so explicitly framed as a journey to develop something like wild pedagogies.

It is not easy to intentionally quiet the chatter in our minds and be receptive to a multitude of more-than-human voices. Anyone who practises meditation understands this. But, part of our task is to try and just be. It is to be open to experience without immediately bringing preconceptions, and a busy-mind analysis, to the situation. Sometimes, in our heart-felt desire to be closer to nature by thinking the issues through so carefully, we trip ourselves up. Nature itself, can evoke a range of emotions that can include fear and anxiety. But with care, it is sometimes enough just to *be*. It can be enough to be close to nature—to embrace our, and indeed its, physicality and sensuousness—and see what arises in that space. And in this, there can be joy in not knowing, in not bringing preconceptions to experience.

In the next chapter we will introduce a series of touchstones that we have found helpful in guiding us through the kinds of conundrums that have been presented in this section. And it is here that we will explain how nature can be a co-teacher, what some of the important tenets of wild pedagogies are, and what questions wild pedagogues might want to be asking themselves and each other as they continue along this challenging, yet necessary, road.

Acknowledgements *Crex crex* is the taxonomical name given to the Corncrake. We have chosen this bird to represent our collective because it was an important collaborator in this project and because its onomatopoeic name beautifully mirrors its call—a raspy crex crex. For some reason, it chooses to fly over England and breeds in Scotland and Ireland. Presumably this is due to loss of habitat in modern England, but perhaps these birds sense some epicenter of empire there? Who is to know?

Notes

1. Sean Blenkinsop, and Marcus Morse, "Saying Yes to Life: The Search for the Rebel Teacher," in *Post-Sustainability and Environmental Education: Remaking Education for the Future*, ed. Bob Jickling and Stephen Sterling (London: Palgrave Macmillan, 2017): 49–61.
2. See for example: Janice Astbury, Stephen Huddart, and Pauline Théoret, "Making the Path as We Walk It: Changing Context and Strategy on Green Street," *Canadian Journal of Environmental Education* 14 (2009): 158–178; Wayne Au, "Teaching Under the New Taylorism: High-Stakes Testing and the Standardization of the 21st Century Curriculum," *Journal of Curriculum Studies* 43, no. 1 (2011): 25–45; Bob Jickling, "Sitting on an Old Grey Stone: Meditations on Emotional Understanding," in *Fields of Green: Restorying Culture, Environment, and Education*, ed. Marcia McKenzie, Paul Hart, Heesoon Bai, and Bob Jickling (Cresskill, NJ: Hampton Press, 2009): 163–173; Bob Jickling, "Self-Willed Learning: Experiments in Wild Pedagogy," *Cultural Studies of Science Education* 10, no. 1 (2015): 149–161; and William C. Smith, ed., *The Global Testing Culture: Shaping Educational Policy, Perceptions, and Practice* (Oxford: Symposium Books, 2016).
3. Arjen E.J. Wals, "What You Can't Measure Still Exists," *The Environmental Communicator* 12 (1990): 12.
4. Elliot Eisner, *The Educational Imagination*. 2nd ed. (New York: Macmillan, 1985).
5. Ibid., 121.
6. Herbert Marcuse, *One-Dimensional Man* (Boston: Beacon Press, 1964): 57.
7. Winona LaDuke, *All Our Relations: Native Struggles for Land and Life* (Boston: South End Press, 2008).

References

Astbury, J., S. Huddart, and P. Théoret. "Making the Path as We Walk It: Changing Context and Strategy on Green Street." *Canadian Journal of Environmental Education* 14 (2009): 158–178.

Au, W. "Teaching Under the New Taylorism: High-Stakes Testing and the Standardization of the 21st Century Curriculum." *Journal of Curriculum Studies* 43, no. 1 (2011): 25–45.

Blenkinsop, S, and M. Morse. "Saying Yes to Life: The Search for the Rebel Teacher." In *Post-Sustainability and Environmental Education: Remaking Education for the Future*, ed. Bob Jickling and Stephen Sterling, 49–61. London: Palgrave Macmillan, 2017.

Eisner, E. *The Educational Imagination*. 2nd ed. New York: Macmillan, 1985.

Jickling, B. "Self-Willed Learning: Experiments in Wild Pedagogy." *Cultural Studies of Science Education* 10, no. 1 (2015): 149–161.

———. "Sitting on an Old Grey Stone: Meditations on Emotional Understanding." In *Fields of Green: Restorying Culture, Environment, and Education*, ed. Marcia McKenzie, Paul Hart, Heesoon Bai, and Bob Jickling, 163–173. Cresskill, NJ: Hampton Press, 2009.

LaDuke, W. *All Our Relations: Native Struggles for Land and Life*. Boston: South End Press, 2008.

Marcuse, H. *One-Dimensional Man*. Boston: Beacon Press, 1964.

Smith, W., ed. *The Global Testing Culture: Shaping Educational Policy, Perceptions, and Practice*. Oxford: Symposium Books, 2016.

Wals, A. "What You Can't Measure Still Exists." *The Environmental Communicator* 12, no. 1 (1990).

5

Six Touchstones for Wild Pedagogies in Practice

The Crex Crex Collective

Abstract The touchstones presented in this chapter are intended to help sustain the work of wild pedagogues. They stand as reminders of what educators are trying to do. And they challenge us to continue the work. These touchstones are offered to all educators who are ready to expand their horizons, and are curious about the potential of wild pedagogies. The touchstones can become points of departure and places to return to. It is suggested that they be read, responded to, and revised as part of an evolving, vital, situated, and lived practice. As such, these initial touch-

---

The Crex Crex Collective includes: Hebrides, I., Independent Scholar; Ramsey Affifi, University of Edinburgh; Sean Blenkinsop, Simon Fraser University; Hans Gelter, Guide Natura & Luleå, University of Technology; Douglas Gilbert, Trees for Life; Joyce Gilbert, Trees for Life; Ruth Irwin, Independent Scholar; Aage Jensen, Nord University; Bob Jickling, Lakehead University; Polly Knowlton Cockett, University of Calgary; Marcus Morse, La Trobe University; Michael De Danann Sitka-Sage, Simon Fraser University; Stephen Sterling, University of Plymouth; Nora Timmerman, Northern Arizona University; and Andrea Welz, Sault College.

Sean Blenkinsop (sblenkin@sfu.ca) is the corresponding author.

S. Blenkinsop (✉)
Faculty of Education, Simon Fraser University, Vancouver, BC, Canada
e-mail: sblenkin@sfu.ca

© The Author(s) 2018
B. Jickling et al. (eds.), *Wild Pedagogies*, Palgrave Studies in Educational Futures, https://doi.org/10.1007/978-3-319-90176-3_5

stones are not intended to be dogmatic, but simply a best gathering of ideas and practices at this time. These preliminary touchstones are, thus, intended to assist educators in practicing wild pedagogies.

Keywords Co-teacher • Curriculum • Education • Environmental • Nature • Touchstone • Wild

Making Landfall and Touching Stone

The islands of the Inner Hebrides are a geologist's paradise, drawing experts and rock hounds from across the globe. Within a small area, there are multiple epic tales of the earth's formation being told. Pre-Cambrian tectonic plates have collided, rebounded, and collided again, crumpling sea beds, exposing the hidden and cracking the surface which in turn has allowed volcanoes to form and lava to flow. And then there has been the ice, entire ages of it covering the region, causing seas to rise and fall while scraping and sculpting the surface and leaving behind erratics, moraines, and notes carved deeply into the bedrock. For those literate in the language of deep time and slow history, the landscape is storied. It is in these glacial notes, these beaches sprung from the depths when the ice melted and the land rose up, it is in these basalt sculptures, and it is in these rolling ancient hills. These stories, like any good epic, are filled with creation, destruction, change, strong emotions, and on-going possibility. For our small group, each time we came ashore, stepping directly into the narrative, the great geostory, and literally "touching stone," we entered a seemingly unique part of the tale. Each island has its own geology but each is also a necessary part of a larger whole, the tale told by Islands of Hebrides. Like any story, there are possibilities afforded and limits implied once pen has touched paper and the copy is printed. Thus, the following thoughts, suggestions, and questions that we have called "touchstones" are limited, too.

Our hope for the touchstones in this chapter is that they sustain the work of wild pedagogues, to be held, and returned to over and over, for guidance, reference, and support. They stand as reminders of what we, as wild pedagogues, are trying to do. And they challenge us to continue the work. We offer these touchstones, to all educators who are ready to expand their horizons, and are curious about the potential of wild pedagogies. Touchstones can become points of departure and places to return to. We suggest reading, responding, and revising them as part of an evolving, situated, and lived practice. As such, these initial touchstones are not intended to be dogmatic, but simply our best gathering of ideas and practices at this time. We hope that these preliminary touchstones can assist us all in practicing wild pedagogies. We welcome further discussion, research, critique, and practice-based elaboration.

In presenting this work, each touchstone begins with a short geological observation drawn from the places we visited. This, in a small way, sets the scene for the reader, it gives a flavour for what we encountered on our journey. This observation is followed by a short "we believe" statement that summarizes the touchstone and serves as a quick reminder for the harried wild pedagogue. The bulk of each touchstone is then comprised of explanatory text with short vignette intrusions that also draw from our experiences in Scotland. The vignettes attempt to bring the natural world actively into the touchstones, and, in different ways, offer a wider range of voices than is usually possible with regular prose. The final section of each touchstone is a series of questions that we hope can prompt readers as they go about practicing and developing wild pedagogies.

Touchstone #1: Nature as Co-Teacher

Passing the Treshnish Isles, raised beaches and wave cut terraces belie a rebounding and resilient crust relieved of the thick weight of the icy Pleistocene. These open aerie platforms, that outstripped the rising seas as a result of melting glaciers, are now carpeted with bluebells, more recent ecological memories of former woodlands.

* * *

We believe that education is richer, for all involved, if the natural world and the many denizens that co-constitute places, are actively engaged with, listened to, and taken seriously as part of the educative process.

* * *

This touchstone reminds educators to acknowledge, and then act, on the idea that those teachers capable of working with, caring for, and challenging student learning include more-than-human beings. This implies more than simply learning from the natural world; it includes learning with and through it as well; and thus, its myriad beings become active, fellow pedagogues. We acknowledge that this can be a challenge for educators "marinated" in a modernist worldview. Yet, we recognize that there can be tremendous benefits to questioning the idea that a single human teacher should be at the centre of teaching and learning, and to expand consideration of what and who an educator is and might be.

* * *

Finding my sea legs … we are three days at sea and I am no longer slipping and flopping around the deck like a fish recently pulled from the water. My body has begun to both respond and anticipate the ship's movement in reply to the waves and wind. I am, like an infant, beginning to enter into dialogue with the world around me. Intriguingly, this has not only taught me some important things about walking, the bruises attesting to a tough mentor, and movement but it also appears to be influencing my thinking. My spoken metaphors are more fluid in nature and I am beginning to read particular waves and stretches of water and understand their meanings in my own context of trying to remain standing and move about the ship.

* * *

Image 5.1 Bluebells blooming. Photo credit: Hansi Gelter

Consider how the field of bluebells described in the opening vignette of this touchstone are telling us where forests once were, and, how the "floating beaches" found 100s of feet away from Treshnish's current coastline tell the tale of lands rebounding from the weight of now melted glaciers. All of these—waves, flowers, and orphaned coastlines—are participating in the process of our coming to know the world and ourselves in it. What can these observations tell us about nature as a co-teacher?

If we take seriously the notion that the natural world is not made up of inert entities; but, rather, it is filled with active, self-directing, and vibrant participants, then our attention towards the affordances of place-based education changes. In seeking to teach *with* nature, educators become open and available to the range of facts, knowings, and understandings that places have to offer. Such attention involves carefully listening to available voices and building partnerships with seashores and forest dwellers. And it will, at times, involve actively de-centring the taken-for-granted human voice and re-centring more-than-human voices.

Such re-centring of more-than-human voices requires openness to the educational opportunities arising in places. For example, a young robin landing on a branch, only to have it snap and fall under its weight, is learning about itself and the nature of a tree branch. But if this is witnessed by us, the experience can become the grist for learning. It can lead to a quick lesson on trial and error, a conversation and future exploration into flight, or an opening into humour—what is funny and for whom. If we take seriously the role of co-teacher, we need to be attentive to moments when our fellow co-teachers are engaging students meaningfully. We need to acknowledge these co-teachers might be offering something more and something different—something beyond our ability. And when these moments arise, we need to provide time and space for the lessons to run their course.

In acknowledging this touchstone, educators are encouraged to encounter the natural world, and its members, in non-hierarchical and equitable ways. This may pose an on-going challenge in conventional education settings. For example, it can mean spending more time outdoors, pushing back against tightly scripted timelines, and changing the contents of classrooms. And it will be an on-going challenge with respect to pedagogical approaches. Here, it can mean less human teacher voice, more independent and place interactive time, and significant changes in the relationships between and amongst natural beings, students, teachers, and subject matters. Changing relationships within educational settings—and also within the schools, communities, and systems from which those same settings arise—will require some radical shifts that are not easily recognized in mainstream settings.

* * *

We are standing on the walls of a ruined thirteenth century luxury home, an armed outpost overlooking Ardtornish Bay, when one of our party recognizes the cry of a stressed herring gull. Shifting focus in the vast blue (yes, it was a sunny day in Scotland) and finding our sky eyes, we locate the disturbed bird and there above it circling and closing in is a huge white-tailed eagle.

* * *

When educators and students attend to the particulars of their places, as in the vignette, they can begin to acknowledge and accept pedagogical invitations offered by the natural world. And at the same recognize affordances and pedagogical possibilities that exist throughout their locales. Many rich conversations about life and death, evolution, binocular vision, self-protection, and working in community can be started as we observe the drama playing out around us. It is important, however, that human learners recognize that these experiences cannot be completely encapsulated by the human imagination. We do not fully understand the interaction between gull and eagle, or bird and sky, or even sun and sight. It is important to be aware that the natural world is not simply an educational opportunity arranged for humans; it is not there just to be picked through by the thoughtful human teacher for the sole benefit of a particular group of students. This awareness requires humility on the part of the human educator, both because human knowledge is now understood as being necessarily incomplete and because sometimes the co-teacher is not just a support teacher, but will take the lead and the learning might go in unexpected directions.

For most human co-educators this touchstone will be a substantive demand, involving re-thinking the concept and role of the educator and reflection upon curriculum and practice. It will require noticing, naming, and even changing metaphors, traditions, and systems that have tended to shut out or devalue the natural world.

With this discussion as background, educators might want to consider questions such as:

- **What habits in my teaching do I tend to fall into that can place at a distance, background, undervalue, or denigrate the natural world?**
- **How can I invite the natural world to be present as a co-teacher in my practice?**
- **How might we as a class contribute to the potential flourishing of each other and those that live lives in proximity to our own?**
- **How have we been able to learn about, with, through, and from members of the natural world?**
- **And, how might we be able to make space for other teaching voices to be heard in their own ways?**

Touchstone #2: Complexity, the Unknown, and Spontaneity

Poured by Vulcan, Staffa, the island of pillars and Fingal's Cave, broods and waits. Spectacular hexagonal basalt columns rise from the sea as a testament to slow cooling and the power of deep cracking. Equally amazing as these vertical columns are the waves and curls of their discarded relations strewn in a multitude of "woodpiles," a sculpted memory of a long completed thermal tai chi, the surprise of witnessing the rhythms and ballets of solid rock.

* * *

We believe that education is richer for all involved, if there is room left for surprise. If no single teacher or learner can know all about anything, then there always remains the possibility for the unexpected connection to be made, the unplanned event to occur, and the simple explanation to become more complex. Knowledge, if given space, is wondrously dynamic.

* * *

For the most part, education is now conceptualised as the transference of a canonized body of knowledge from those who know to those who do not. What is considered knowledge, and what is worth knowing, is largely predetermined by those in control, And yet, even in this carefully constructed space, there lurk shadows and forgotten strands of complex interconnections. Wild pedagogies seek to open up possibilities for embracing complexity and spontaneity in ways that imply re-negotiating educational practices. Embracing complexity will require encounters with that "which cannot be known," which cannot be predetermined and prescribed in advance. Complexity can be understood as dynamic, fluid, and unpredictable, and is best described in reference to qualities without fixed boundaries. It stands in contrast to a static, deterministic, and linear view of the world. As Noel Gough suggests, "complexity invites us to understand our physical and social worlds as open, recursive, organic, nonlinear and emergent, and to be cautious of complying with models and trends in education that assume linear thinking, control and predictability.[1]" This implies that educators need to, at least in part, relinquish the control and self-domesticating forces that are ingrained in our pedagogical thinking and practices. And it implies that they will need to be more open to spontaneous, and sometimes surprising, occurrences.

* * *

The sea has gone quiet. After yesterday's tempest, this placid, slate-grey surface seems to be asking us to forget the storm, but that is impossible. As we climb down the precarious ladder tied to our sailboat's hull and settle ourselves in the tiny unreliably motored dinghy for the 300 m ride to shore, we are quiet humbly remembering yesterday's power display and our own limitations in the face of such violence. And yet, it is this ecozone, this risky space between 82 acres of basalt monolith that is the island of Staffa and 100 feet of sailing vessel where the seabirds really make themselves known. We are surrounded by rafts of guillemots and razorbills, while above us flying in huge circular patterns from cliff homes to sea feeding and back, are hundreds of boisterous puffins. These avian wonders and their expressive cacophony tell us clearly of the abundance of this interstitial zone between safe havens.

* * *

Image 5.2 Basalt columns on Staffa. Photo credit: Hansi Gelter

Mainstream education often seeks to position itself in the safe places, the solid ground of land and boat, and yet it is often in those interstitial spaces where productivity and possibility exist. Wild pedagogies challenges ideas of control in education by embracing complexity, inviting risk, and allowing for emergence. For example, what might spontaneously arise, unpredictably and unplanned, from interactions between learners and nature?

First, complexity challenges our notions of linear and reductive understanding as it underpins many current educational systems and practices. And second, leaving space for the unknown and spontaneity is a way to respond to dominating tendencies of educational control. For example, universal and measurable standards are created based on a set of concrete truths; schools function to define and legitimize the places in which learning can occur; students are managed via timed programming, normalized instructional locations, and prescribed outcomes; and knowledge is understood to be amenable to fragmentation and deliverable in parts independent of the context from which it arises. But maybe there is something important going on in that zone between the comfort of the boat and the solidity of the land, something unpredictable that can arise from this interstitial space.

This second suggested touchstone for wild pedagogies involves actively embracing the unknown, complexity and emergence, allowing space for the spontaneous. All three of these components involve a kind of pushing off from the safe centre, an undoing of the human as centre of the world, as managerial arbiter of everything, in order to allow other ideas, possibilities, spaces, beings, and imaginations to emerge.

* * *

We are coming to the end of our first full day together and we have gathered in the saloon as a group in order to try and put the finishing touches on our proposed community norms. This facilitated process of trying to determine who we want to be as a group and how we want to interact with one another is a classic activity, but our process seems to be grinding to a halt.

People are tired. It has been a busy day of meeting, greeting, and dealing with the whys and wherefores of the boat, this colloquium, and more. Suddenly there is a wild clatter on the deck above us. The voices of several raucous gulls breach our gathering. The excitement draws our attention and yet hesitation. Some wish to respond, some seek to stay with the current work, and some are caught in-between. But the disruption continues and there is a rush deck-side to catch three gulls in the throes of a dispute over a bit of food. What is striking is that in the dispute much of the food is lost and nobody really benefits in the end. Everything quietens down in a few minutes and a solitary gull establishes itself high in our rigging while the others fly off in search of new plunder. Our little group of humans remains on deck, quietly, our impasse disappears and we unanimously agree to change our norms such that we will seek to actively include the natural world as much as possible.

* * *

The spontaneous nature of an encounter with hungry gulls reveals educational possibilities. The gulls were a clarion reminder that the world exists; and it was good listening to them. In some ways, our own ideas, limitations, and impasse were played out in front of us. Active intrusions can lead to powerful "aha" moments. Just as positions and perceptions were overturned by this encounter, wild pedagogies might allow for spontaneous encounters that challenge implicit ways of knowing and being, and even disrupt cultural norms. For us, our preferred norms were literally rewritten as a result.

Viewing and understanding the world as complex, spontaneous, and mysterious is difficult. For educators this means climbing down the outside of the sailboat and sometimes puttering out onto the giant unknown in a somewhat unreliable dinghy. This also means resisting the urge to grab the textbook, to offer the "right" and simple answer or do what has always been done. This resistance to the solid and controllable might require questioning current metaphors, practices, and understandings of what it means to learn and to know. It also involves overcoming mainstream education's reliance on defined outcomes, known standards, and measured results. There must be more room for learning that is fluid, flexible, and diverse.

Educators might want to consider questions such as:

- What might I do to embrace complexity in my teaching today?
- How was I able to empower learners to journey into the complexity of knowledge and not reach for the easy, seemingly final, answer?
- How did my practice today take risks in moving away from the full control of assumed ends? And how might it continue with that tomorrow?
- Is there room for the unknown, spontaneous, and unexpected to appear and be taken seriously in our educational work?
- Did learners encounter the incomplete nature of knowledge today?

Touchstone #3: Locating the Wild

Awe inspiring insights into paleo-climates and unsorted Precambrian glacial tillites create a ragged western wall protecting early monks from the worst of the sea's rages while they gently and tenderly settled in this wild place. Their rudimentary accommodation and places of worship are now in ruins, though not so the objects of their devotions enduring gaze or reasons for choosing this site.

* * *

We believe that the wild can be found everywhere, but that this recognition and the work of finding the wild is not necessarily easy. The wild can be occluded, made hard to see, by cultural tools, by the colonial orientation of those doing the encountering, and, in urban spaces, by concrete itself.

* * *

Six Touchstones for Wild Pedagogies in Practice

The Norwegian eco-philosopher, Arne Næss, was once asked about what teachers can do in urban areas, and how teachers might meet some of the challenges in taking children outside. He replied:

> In the schoolyard itself, you find a corner where there is just one little flower. You bend down—you use your body language—and you say: "Look here." And some answer: "There is nothing there." And then you talk a little about what you see: "This flower here, it's not the season for it. How can it be there this late in the year? And look at it. It certainly has need of a little more water; it's bending, look at the way it bends. What do you see when it's bending like this?" I call teachers who behave like this "nature gurus." It is a little more like an Eastern kind of education. More in terms of personal relations. Try to make them see things they haven't seen before. Use your body language. And even inside the schoolyards you find nature's greatness.[2]

As Næss suggests, there is potential to encounter the wild in a range of settings. Given that the growing majority of us live in super-urban, urban, and suburban places where the wild may not be easily and immediately apparent this touchstone presents both fertile ground and difficult work. In bringing students to encounter the wild there are no educational guarantees: there is no simple solution to the problem of how to facilitate students' encounters with the wild, the self-willed, and self-arising others that surround us. There is no simple way to nourish that curling wiggling, reverberating, upending version of wildness that exists within us.

* * *

It is a beautiful spring day in the Hebrides. Here and there we can see yellow irises just coming into magnificence. There is lush grass in the low-lying fields and in the distance, beyond white sand beaches that remind us of those in the Caribbean, the sea sparkles. We are surrounded by the burbling and murmur of waters on the move, shore birds creating homes, and small trees coming into full leaf. But, as someone has just pointed out, our group of humans has turned into a noisy gaggle. Attention drawn to each other's voices, to the lines of discussion connecting mouth to ear. Our focus has turned inwards, as it so easily does in this world where humans are the centre, and the sun, the colours, the voices of others have slid into the background yet again. We have, in spite of ourselves, created a human "social bubble."

* * *

Many of us have spent days facilitating groups in remote areas only to have students turn away from encounters with the wild. We have witnessed them retreating into the comfort of their tents, social groups, or their own intellectual voyages and other unconscious habits. On the other hand, many have also experienced moments, immersed in deepest urban jungle, when a single green sprout has leaked through the pavement, and punctured the human ecology. By raising a leafy resistance to the "power of humans" story, a wild seed can be planted in an unsettled student. The wild is everywhere, and yet we also note that encountering it often appears to be easier in wilder, more self-willed places. One wonders if the poetry of those early monks living on wild, ragged, and isolated isles would have been different if they had done their writing in the quiet of a Dublin monastery. What would have happened to their humble odes filled with the voices of corncrakes and cuckoos? Their ideas of God and land? The challenge for many urban-based environmental educators is, then, that the murmur of wild can be overwhelmed by the noise, smell, and dominion of human constructions.[3]

Encountering the wild provides educators with complexity, opportunity, and challenges. The anti-colonial literature of Tunisian scholar Albert Memmi,[4] for example, when read with an eye to the modern human relationship to the natural world, offers a troubling analysis for wild pedagogues. Following Memmi as he develops themes exploring how the colonizer operated in order to oppress the peoples of North Africa, it is not a difficult stretch to see those same themes played out in relation to the natural world. Such an analysis implicates educators in a complex project that is not simply about providing opportunities for students to encounter the wild. It also requires helping students to see their own privilege in light of the destruction that western human-centred culture has wrought on the world. This move to see and respond to the "natural world as colonized," particularly in urban spaces, is complicated by the psychic challenges that all decolonizing work entails. It can help to build students' fortitude so they can stay with, and even welcome, difficult and emotional discussions. Yet, this educational project is challenging because language, ways of being, structures of schools, and urban settings are all oriented in ways that draw students' attention away from the wild, concealing both their alienation from it and their privilege.

Educators might then consider questions such as:

- **How can I make it possible for my students today to have encounters with the wild and/or self-willed communities that inhabit the spaces we are in?**
- **In what ways can I notice, and respond to human-centred and colonizing perspectives that we might encounter today?**
- **What can I do to provide ways for the wild in our encounters to be acknowledged?**
- **How did I notice and overcome my own and my students' colonial habits in relating to the natural world?**
- **How might we recognize and how might we encourage acts of resistance enacted by wild beings? And how can we help students to develop the ability to "lean in to" and even welcome difficult encounters with privilege, alienation, and colonization?**

Touchstone # 4: Time and Practice

Our trip began with a circumnavigation of the island of Mull. A place dominated by layers of lava that have poured out of the earth at different geological times. The constant upheaval has meant that older lava is shot full of intrusions. And the unpredictable nature of flows and cooling has meant that there are dykes and small cliff edges all over the island. Landing on Mull is to dip in and out of deep time, dip in and out of ecological and cultural history.

* * *

We believe that building relationships with the natural world will, like any relationship, take time. We also believe that discipline and practice are essential to this process.

* * *

This touchstone is about process, and we are offering two key components for assisting individuals in the process of change—time and practice.

For our current purposes we are thinking of time in two different ways. The first way is the one we are most familiar with: clock-time. Much of the work described in the previous touchstones requires this kind of time. It is required for building relationships with beings, things, and places. We must spend time—lots of it—immersed in, dialoguing with, and learning with, from, and about the natural world. The second way we are thinking of time suggests a less linear, more organic, deeper form, similar to the one encountered on the island of Mull.

* * *

Approaching the Garvellachs on a breezy spring day is an adventure. There is no landing site so we are forced to scramble onto barnacle-encrusted rocks as the dinghy noses into a small bay. From a distance, the island appears too small and rugged to have served as home for anyone and yet, after scaling the rocky shoreline, small flat enclosures and pocket sized dwelling sites become visible. These run up the slope next to the tiny, permanent, two-litres-per-hour, freshwater stream. Across a low hill we find the best preserved of the monks' cells. A beehive structure built from stone and physics stands 15 feet high. It is a quiet place for a life of discipline, perched precariously and exposed to the vagaries of wind, weather, and wildlife.

* * *

Six Touchstones for Wild Pedagogies in Practice

Image 5.3 The Lady of Avenel in the Garvellachs. Photo credit: Hansi Gelter

Clock time is also required to support learners as they grow into new, wild habits. For pragmatic philosophers, the process of habit change is understood as one of deep examination of self and culture. In light of this, learners must come to recognize that they have already developed many habits to enable them to navigate the existing cultural waters of their time and place. Educational moments may arise over the course of a wild curriculum that offer opportunities for learners to recognize that such habits exist and to open them to critique and revision thus making it possible to engage in a process of self-creation where learners begin to enact new ways of being in the world. However, this process is challenging and clock-time consuming. Previous habits are resilient and embedded, well-practiced response mechanisms used for navigating the world.

One wonders what habits the monks were forced to recognize and change as result of living on these tiny islands with very little fresh water. And how were these habit changes linked to deeper discoveries as a result of living in these isolated and exposed islands? Were they inspired by these places to change their regard for the natural world and their own place in it?

By contrast the second way of looking at time involves seeing it as an organic and living process rather than a linear steady tick-tock. Sometimes it is called "deep time." Learners come to recognize that they are continuously having new and different experiences that appear at odds with each other, that are on-going and incomplete, that are complex and uncontainable. Reason is of limited usefulness. Sometimes the world around seems much larger than we can comprehend. It was like this for us when, crouching on exposed shores of the Gravellachs, we tried to imagine sixth century monks inhabiting that place, or when encountering the ancient stone on Mull, we contemplated its age. Here intuition, a product of deep time, plays a more important role than reason, which is a product of more recent cultural history. Finding a place for intuition, sustained by organic time, is needed to allow for more expansive wild encounters. And it will be needed to tell a different, renegotiated, geostory. Acquiring intuition, and a relationship to organic time, will require discipline, even practice, and will only be learned over time. Yet it is important to wild pedagogues and can lead to what Arne Næss might call a deeper and bigger self.

* * *

We are meeting, on this beautiful calm evening, in the forward saloon of the sailboat. It is where we, as a group, seem to do most of our living. We eat here, we gather for meetings and discussions, and we often just hang out in this comfortable space. Low windows surround us on three sides and from a slouched position I can see out to the darkening sky. The discussion this evening is slow and somewhat desultory. We have had a full day and our bunks are calling. Then all of a sudden our lethargy is interrupted by a chirrup from an unexpected visitor and a shiver of excitement runs through the group. Quickly and quietly we gather on the outer decks to watch an otter frolic and even engage with its audience. Five minutes later it disappears from sight, but its input guides our discussion for another 90 minutes, as the wind lightens and the sun sinks into the sea.

* * *

Closely associated with time are invitations to practice. The first invites teachers to develop their own practice in a way that deepens relationships with local places and beings. This work, sometimes likened to meditative practice, requires listening more deeply and becoming attuned to more-than-human co-teachers. It can be a first step leading towards a radical reworking of relationships. This form of practice, like all others, suggests on-going work. Environmental relationships that blossomed in childhood, or that occurred in short time bursts in the backcountry, while at the cottage, or at an outdoor learning centre are often important starting points, but insufficient to be considered a practice unless they have been developed and sustained. Wild pedagogies, as a practice, requires the kind of on-going attention and discipline that any other practice might entail.

Our second invitation to practice involves teachers developing the will and ability to rework their own pedagogies. Consider the earlier vignette of the distressed seagull and the hungry eagle. In this case, someone needed to have an acquired practice with the natural world in order to recognize that particular vocalization of the gull. This educational practice first allowed one of our companions to draw attention to the "noise-maker" and, second, to follow the lead of the noisy co-teacher up to the eagle. Then he shared that interaction with the rest of us while finding ways to respond to the different reactions of the group members, the learners took. He also needed the intuition to make use of this particular event at this particular time. The success of our group experience rested on our teacher's lifetime of attentiveness.

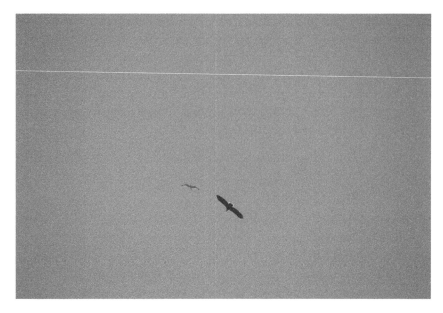

Image 5.4 Voices amongst us. Photo credit: Hansi Gelter

Developing new practices will require reflection, risk taking, experimenting with possibilities, examining successes and failures, and then repeating this process over and over. Students and teachers need opportunities to practice. Wild pedagogues and their students will also need time in that interstitial space between old habits of practice and traditional relationships with the natural world, and new radically different ones.

Educators might want to consider questions such as:

- **Can I leave enough space in my teaching to allow my students and myself to engage with natural places and with beings that are nearby?**
- **Can I recognize that we often go at different speeds, and that some will need more time than others?**
- **How are we, together, able to find ways to step out of the linear time of the school system and encounter time working in different ways?**

- Was I able to notice, respond to, and support students who were trying out new habits?
- How can I maintain and nurture my own practice of immersing in and building relationships with the places and beings I encounter?
- Am I noticing my practice, trying new things, reflecting on what has been attempted?
- Were there opportunities for my students to develop their intuition?

Touchstone #5: Socio-Cultural Change

Stepping ashore on Iona takes us deep into the Archean gneisses at the core of the country. These appear where the earth has been moved alongside ancient schists. Sprinkled across the surface are Silurian and Devonian granites brought much later as glacial erratics. We find all these rocks carefully positioned in the islanders' stonewalls and the abbey buildings celebrating St. Columba.

* * *

We believe that the way many humans currently exist on the planet needs changing, that this change is required at the cultural level, and that education has an important role to play in this project of cultural change. We also believe that education is always a political act, and we see wild pedagogues embracing the role of activists as thoughtfully as they can. Current norms of the dominant Western culture, many of which infuse mainstream education, are environmentally problematic. In response, we seek wild pedagogies that are actively and politically aimed towards telling a new geostory of a world in which all beings can flourish.

* * *

We have seen plenty of evidence that the Earth is in the throes of geological change. The stable climate and diverse eco-systems of the last epoch are rapidly changing. As educators we have always aspired to play an important role in preparing students for the future. But what happens when the future is no longer predictable? This touchstone seeks to respond to this question by first encouraging wild pedagogues to challenge the dominant versions of education that serve to confirm the *status quo*. The current educational "normal" is dangerously replicating the kinds of society and culture that enabled this geologic change and loss of bio-diversity in the first place. Disrupting current trends and preparing learners for an unclear and virtually unknown future requires a conscious shifting of values and educational priorities that is fundamentally political in its purpose and practice.

* * *

There is an interesting cultural clash starting to appear on the boat. One group more informed about Scotland and the social realities of crofters and clearances, has a more complete agenda for things to see and important places to visit to further the discussions we are having. As a result of limited and directionally troublesome wind this group is suggesting more time spent under motor. The other group, folks who were less privy to the original organizing and who are currently intrigued by the ocean landscapes and its movements, has been pushing for more time under sail, even without wind, released from the demands of wind, current, and agenda. For the first group this desire to float is a kind of classic forgetting that human histories are part and parcel of landscapes, and that behind their agenda lies the history of colonialism and human intervention. For the latter group, the no-motor advocates, the first group seems to be falling into a common "be productive" myth, needing to be doing rather than recognizing the importance of not-doing. They worry that the motor is obscuring the water world around us and setting at a distance our relations to the natural world.

* * *

Wild pedagogies call upon educators to engage in these complex, contentious, and challenging situations, as described in the vignette above. It requires dialogue that brings out a deeper motivation and belief in learners, and that they consciously embrace their role as political beings. Acknowledging oneself as a political being or activist means recognizing that the choices we make are negotiations of power. Such a stance has often been seen as anathema to the professional educator. While it is widely acknowledged that education is value-laden, how to respond to this realization remains contested. There will always be disagreement around the introduction of politics into education and the danger of its politicization, particularly in an era of ideologues and demagoguery. Educators rightly fear teaching that imposes their own political or religious views onto students in ways that affect students' ability to choose for themselves in a secular, inclusive state.

However, it is clear that any choice made in classrooms by teachers is political and has implications for the world that is being brought to the students, to paraphrase philosopher Martin Buber.[5] To teach to a supposedly impartial curriculum is to acculturate the students to a *status quo*, a specific paradigm, and a set of beliefs and practices, those of the dominant culture. Such an unquestioned and unexamined curriculum presents the student with the prevailing political norms and, since both students and educators are immersed in that same cultural and social environment, the political message goes unrecognized. This invisibility and normality makes *status quo* politics extremely potent. It is, after all, the daily bread of the society.

If an educator's work is to teach about the world, then she must help students to see, and name, the politics that shape our collective norms and relationships. Teachers can remain unpartisan, but they cannot remain politically neutral in their work. We are exercising power and positing value in the choices of: what content to share and what pedagogy to use, which ideas will be followed up and deemed worthwhile, which power dynamics are condoned or disrupted in our classes, and which histories are told. Despite the plethora of prescribed curricula, policies, and bureaucracy that can constrain educators' options for content and pedagogy, our work can still be a force for cultural change.

Thus, wild pedagogies are explicitly and deliberately about enabling mutually desirable socio-cultural change. In choosing content and pedagogies there is an aim—humbly submitted—for this work to change the relationship humans currently have with the natural world. We hope for a relationship that is much more equitable and interactive, that pursues flourishing for all beings for the express purpose of stopping the massive destruction being wrought and to mitigate accompanying problems such as climate change.

Where does the educator fit in to this type of transformation? Does the educator herself act strategically to influence the dominant power structures? We answer "yes." Our actions always influence power structures, but often subconsciously. When we are unaware of this influence, we are in danger of aligning with the *status quo*. As long as we can acknowledge this, we have options for understanding our own power, how it is currently being used, and how we wish it to be used. Wild pedagogies proposes that we exercise our options with consideration for our relations with other beings, that we tend to the earth and our relationship with it, and that we see the more-than-human world as political actors. In doing all this we will wild our conceptions of socio-cultural change. A small example may help to clarify the point.

* * *

The sails are up and the boat catches the wind. The engine's hum goes silent and the ship leans (or heels) at an angle, starboard side sinking into the sea. The dominant presences now shaping our lives are water and wind. They set the agenda, continuously breathing into the ship, into our balance of legs and feet, hearts, and minds. Intriguingly, with the sails up, the boat has become a new conduit for a different kind of communication. We are offered a new way of understanding and encountering air and wave, each playing with the other. Our concepts and understanding are being shaped by the sensory organ that is the boat itself. The wind is translated by the waves into the hull of the boat which in turn translates that into our bodies and we become interpreters of sea and sky.

* * *

This touchstone is genuinely wild, for we cannot know in advance what the outcomes will be or how future learners will enact their learning. We are, in some ways, under sail without a clear sense of the destination and much to still learn. What is intended is that wild pedagogies will change education, how it is conceived and enacted, and this will disrupt the invisible politics of the *status quo*. It may even result in learners who are more loving and compassionate in deploying their power, more vigilant and careful in the values they support, and who consider themselves players in healing or resisting, restoring or creating, doing or not-doing as the situation requires.

Educators might want to consider questions such as:

- **How does my practice respond to the existing curriculum and values that are embedded in my workplace? Am I satisfied with my response? Why? What are my criteria for satisfaction?**
- **Were there situations that arose in my lessons that allowed learners to consider their current relationships with the natural, and did they have the right to change them?**
- **What politics of the natural world did we encounter and how was that brought into our work?**
- **Where are my habitual ways of doing things still limiting possibilities?**
- **What possibilities for future scenarios am I raising with my students? Am I offering realistic tools to support proposed change?**
- **What am I doing to help students develop political agency?**
- **What human and more-than-human voices are included and/or excluded in the histories, explanations, and readings that I share with my students?**

Touchstone #6: Building Alliances and the Human Community

Our trip ends as we reach the Easdale slates of Luing and the windswept, dark grey, relict quarries of a once thriving worldwide trade. Open pits are scattered everywhere, homes built from the stone that was at hand, and soli-

tary sheep grazing on the open hillsides all act as testament to the intrepid and insidious nature of the resourceful beings who carved out life here.

* * *

We believe that the colonial ethos of resource extraction is not separate from, but is yet another shade of the many hierarchies of dominance that exist amongst humans. For this reason wild pedagogues seek alliances and build community with others not only in the environmental world but across all people and groups concerned with justice.

* * *

This touchstone acts as a reminder to wild pedagogues of their interdependency with communities and considers their presence and role in wild pedagogies. Community is most often thought of in human terms, and there is good reason for this. Like many creatures on earth, humans depend on social relationships with their own kind for love, support, and protection. All humans, whether they are headed into wilder areas or creating the best curricular opportunities they can are nodes in networks of rich and reproductive human labour. Without communities of people and more-than-humans caring for us and enabling our lives we are incapable of even beginning to act. And yet, the ways in which we humans interact with one another can also be more or less wild. It is not difficult to imagine social scenarios where wildness and freedom appear amongst us. Educators interested in wild pedagogies might well seek those wild interstitial zones just as we humans might seek scenes of healthy wildness in our communities where people can express themselves freely and not feel stifled. Although we may feel uncertain and intimidated when first entering such places, there is also a sense of liberation and comradeship that we experience. Groups that come together in this way have more autonomy and a greater sense of empowerment, as do the individuals involved in them. It is a "wild" process, characterized by the equitable sharing of all voices and a need for decisions to be made collectively by all concerned. Like wild pedagogy, such spontaneous democracy is concerned with the notion of "will;" it is about the self-will of the group or of the community, and an ever-present struggle to identify and make decisions according to that will as equitably as possible. In the context of wild pedagogy, democracy of this type helps us remember that there are communities, made up of humans and more-than-humans, affected by all decisions, and that all involved ought to have a say, in whatever language, voice, and form is their own.

* * *

We have anchored in a shallow bay for the evening and the boat is pulling gently on her long chain. The wind has been slowly shifting directions for the past few hours and we have swung from east to west. The sky is trending from radiant blue to resplendent black as the sun hides behind the flank of the nearest island. The few clouds scattered in the sky are touched with colour as yellows slide to oranges and then fade into reds. There is a restless creative energy in the group. We are nearing the end of the trip, have encountered, thought about, and explored much. We have been immersed in a wild scape and the Inner Hebrides is rubbing off on us. From somewhere in the ship, a guitar appears, and then another. Soon the cabin is filled with music, there is dancing and eclectic movements, there are invented instruments (e.g. spoons on tables, fingers on glasses, cardboard whiskey containers on resonant foreheads), and there are voices coming together and being proffered out. This small human community is becoming wild together, turning in and turning out. And in the distance, the seals howl an answer.

* * *

But as much as community is everywhere, it can often be forgotten or neglected in a culture that is predominantly individualistic. Hence, the first suggestion this touchstone makes is that educators identify the multiple communities of which they are a part, and the complex interdependent composition of those communities that always implicates the more-than-human. Once identified, educators are better able to recognize and maintain those that are positive, attend to and heal those that are troubled, and even remove themselves and their students from those that are limiting and destructive. And in tending to our communities in these ways, educators expand the possibilities for their learners to understand themselves as belonging to and being responsible for others.

A flourishing, wild community is one that sustains its members, allows them to flourish. They also challenge us in important ways by helping us to become different, potentially better. When we have hard and uncomfortable work to do, communities can be positive spaces to simultaneously encourage and challenge us. They support and help us to do the work while reminding us of our commitments. In this way, communities help build individual and collective resilience. Healthy communities are places where people can take risks, where we can try out new ideas or practices, where we can depart from the *status quo*. People find belonging, friendship, and joy in their communities. We all need supportive communities as we attempt to re-wild our lives, pedagogies, and places where we live.

Communities are also locations where the important work of all the previous touchstones occur. On our own, we are prone to the limitations of our own imagination and often end up recreating the same systems and relationships that we seek to transform. Multiple perspectives allow each of us to see beyond our own limitations. We also recognize that socio-cultural change and issues like climate change and environmental justice is not the work of wild pedagogues alone. We encourage wild pedagogues to identify allies and seek synergies that help respond to the challenges while empowering learners.

* * *

We are moored in a sheltered passage between a small islet and the island of Iona. It is evening, the wind has disappeared and the sky is trending from radiant blue to resplendent black as the sun hides itself behind the low flank of the bigger isle. The group, pleasantly worn out from excursions across Iona, and happily sated after a meal, is spreading out around the ship. Quiet conversations, gentle strains of music, and silent witnessing of the sunset seem like appropriate responses to place and time. Suddenly that changes as a voice in the stern rings out, "dolphins!" A shiver of excitement runs through the boat. To the south and moving rapidly towards us through the passage are several dolphins. At each surfacing the disappearing light reflects deep hues of magenta from their sleek backs and fins. We are entranced. As they come parallel with us, one dolphin explodes out of the water, performs a full somersault, and slides cleanly back into the water with a skill and precision beyond anything I could imagine. The reverberations of its joy and power leave us breathless, as it clearly performed for our benefit. Mixed in and amongst the emotions of joy, wonder, and desire to see more, I am touched by a sense of responsibility as well.

* * *

This touchstone suggests that we can "wild" our communities when we seek collaboration amongst allies. We can be proactive, too, by asking ourselves questions: Who are the people in the community? Are there like-minded people amongst them? What values are shared by the members? How can we find common ground with them? As we answer these questions, we learn from each other about how our concerns are mutually shared and, by carefully working with each other, we can identify areas of contention and seek to resolve the differences. This is part of the work of wild pedagogy. There are no "right" ways to do this work, it evolves. Re-wilding of our communities is about recognizing the agency within all beings, including human beings, and the ways in which that agency has been ignored or oppressed, and then striving for a positive resolution equitable to all, including the more-than-human world.

Educators may want to consider questions such as:

- **Who makes up my communities when I think of doing wild pedagogies work? Why? Who is not included, and then, whom do I want to be included? And why?**
- **How do the various communities I am part of make decisions? Who is impacted by the decisions we make? How are all those affected included in the decision-making process?**
- **What might I do that would bring the natural world more explicitly into community decision-making?**
- **How do I support and make better my communities and how do they support and make me better? How can I bring these questions into my classroom?**

- How may my communities encourage one another to imaginatively depart from the *status quo*? How do we encourage one another, allow for mistakes, and also challenge and push each other?
- How have I worked to build trust within my communities?
- What are the affordances, skills, and possibilities that already exist in my communities, and how might we connect in ways that benefit all?

> **THESE SEA-WORN ROCKS**
>
> These sea-worn rocks will be here long after me
> and you will see them with my eyes
> these black wet rocks will remain
> when we are long gone
> we see them with the eyes of those
> who beached their curraghs on this bay
> and sheltered under these cliffs
> and those who unlocked slate
> to makes roofs and walls
> tonight we gaze in wonder
> at the ceaseless rush of sea on shore
> and you will think of us this night.
> *Norman Bissell*

Acknowledgements *Crex crex* is the taxonomical name given to the Corncrake. We have chosen this bird to represent our collective because it was an important collaborator in this project and because its onomatopoeic name beautifully mirrors its call—a raspy crex crex. For some reason, it chooses to fly over England and breeds in Scotland and Ireland. Presumably this is due to loss of habitat in modern England, but perhaps these birds sense some epicenter of empire there? Who is to know?

Notes

1. Noel Gough, "Towards Deconstructive Nonalignment: A Complexivist View of Curriculum, Teaching and Learning," *South African Journal of Higher Education* 27, no. 5 (2013): 1220.

2. Arne Næss with Bob Jickling, "Deep Ecology and Education: A Conversation with Arne Næss," *Canadian Journal of Environmental Education* 5 (2000): 54.
3. Michael Derby, Laura Piersol, and Sean Blenkinsop, "Refusing to Settle for Pigeons and Parks: Urban Environmental Education in the Age of Neoliberalism," *Environmental Education Research* 21, no. 3 (2015): 378–389.
4. Sean Blenkinsop, Ramsey Affifi, Laura Piersol, and Michael Derby, "Shut-up and Listen: Implications and Possibilities of Albert Memmi's Characteristics of Colonization Upon the 'Natural World,'" *Studies in Philosophy and Education* 36, no. 3 (2017): 348–365.
5. Martin Buber, "Education," In *Between Man and Man*, trans. Reginald Smith (New York: Routledge, 2002): 98–123.

References

Bissell, N. *Slate, Sea and Sky: A Journey from Glasgow to the Isle of Luing*. New ed. Edinburgh: Luath Press, 2015.

Blenkinsop, S., R. Affifi, L. Piersol, and M. Derby. "Shut-up and Listen: Implications and Possibilities of Albert Memmi's Characteristics of Colonization Upon the 'Natural World.'" *Studies in Philosophy and Education* 36, no. 3 (2017): 348–365.

Buber, M. "Education." In *Between Man and Man*, trans. Reginald Smith, 98–123. New York: Routledge, 2002.

Derby, M., L. Piersol, and S. Blenkinsop. "Refusing to Settle for Pigeons and Parks: Urban Environmental Education in the Age of Neoliberalism." *Environmental Education Research* 21, no. 3 (2015): 378–389.

Gough, N. "Towards Deconstructive Nonalignment: A Complexivist View of Curriculum, Teaching and Learning." *South African Journal of Higher Education* 27, no. 5 (2013): 1213–1233.

Næss, A., and B. Jickling. "Deep Ecology and Education: A Conversation with Arne Næss." *Canadian Journal of Environmental Education* 5 (2000): 48–62.

6

Afterwords

The Crex Crex Collective

Abstract In many ways the ideas in this book build towards the touchstones in the previous chapter. In the meantime some educators have already been drawn to wild pedagogies and found the touchstones helpful. Samples of their stories are presented in this chapter, for a couple of reasons. First, these are examples of individuals who are openly re-negotiating their practices, and are actually making changes—they are walking the talk. We think that exemplars, or concrete examples can do work, too. They can help people see themselves through images of others making changes. And they can be inspiring. Second, wild pedagogies

---

The Crex Crex Collective includes: Hebrides, I., Independent Scholar; Ramsey Affifi, University of Edinburgh; Sean Blenkinsop, Simon Fraser University; Hans Gelter, Guide Natura & Luleå, University of Technology; Douglas Gilbert, Trees for Life; Joyce Gilbert, Trees for Life; Ruth Irwin, Independent Scholar; Aage Jensen, Nord University; Bob Jickling, Lakehead University; Polly Knowlton Cockett, University of Calgary; Marcus Morse, La Trobe University; Michael De Danann Sitka-Sage, Simon Fraser University; Stephen Sterling, University of Plymouth; Nora Timmerman, Northern Arizona University; and Andrea Welz, Sault College.

Sean Blenkinsop (sblenkin@sfu.ca) is the corresponding author.

S. Blenkinsop (✉)
Faculty of Education, Simon Fraser University, Vancouver, BC, Canada
e-mail: sblenkin@sfu.ca

© The Author(s) 2018
B. Jickling et al. (eds.), *Wild Pedagogies*, Palgrave Studies in Educational Futures, https://doi.org/10.1007/978-3-319-90176-3_6

are relevant to educators in a wide variety of settings and this chapter has, thus, included stories from a range of educators and educational settings.

Keywords Co-teacher • Curriculum • Education • Environment • Nature • Touchstones • Wild • Re-wild • Pedagogies

In this book we have worked towards bringing ideas and practices together. Like a well functioning ecosystem, when working together, ideas and practices allow each other to flourish. In the end, though, it will be through our practices that our friends, colleagues, learners and, indeed, Earth will know that we have changed our ways of being. As we have attested, to do this will be challenging; and it will not happen all at once.

In many ways the ideas in this book build towards the touchstones in the previous chapter. Highlighted amongst them were suites of questions that educators could ask themselves as a means of challenging and transforming their practices, and re-negotiating their ways of being in the world. The value of these questions will ultimately lie in the work they do. We encourage readers to re-write them and craft new questions, if they aren't doing enough work, or if they aren't heading you in what seems to be the right directions.

In the meantime some folks have already been drawn to wild pedagogies and have found the touchstones helpful. We have included a small sample of their stories below, for a couple of reasons. First, these are examples of individuals who are openly re-negotiating their practices, and are actually making changes—they are walking the talk. We think that exemplars, or concrete examples can do work, too. They can help people see themselves through images of others making changes. And they can be inspiring.

Second, we believe that wild pedagogies are relevant to educators in a wide variety of settings and have, thus, included stories from a range of educators. We want to stress that this book isn't just for those working in schools or universities. The stories that follow arise from: an art gallery; public schools and their administration; a small Scottish non-governmental organisation; a Japanese University professor's experience of her country's move away from traditional knowledge towards a universal and capitalist educational system; private sector early childhood education and a college programme; and academic research that challenges research norms. From this modest beginning, we expect the breadth and depth of exemplars to build. For now, this is a beginning.

Out the Gallery Doors

Vivian Wood-Alexander is an educator at the Thunder Bay Art Gallery in Canada. She first encountered wild pedagogies, as an idea, during a graduate course at Lakehead University in 2012. She also participated in 2014 *Wild Pedagogies: A Floating Colloquium*, in the Yukon, There she revelled in the artistic possibilities of river clay as co-artist and co-teacher.

* * *

At the entrance of the public art gallery where I work, I push open the heavy glassed doors. A class has arrived. I look at the line of children, and then over their heads at the trees and sky behind them, visible through the space in the familiar caribou sculpture by Ahmoo Angeconeb—a stylized steel caribou, totem of the creative ones. In that moment wild pedagogy arises for me. Instead of inviting them in, I step out toward them, I am unsure of where to begin but I have decided that their first creative experience will not be inside. The wild need not always be located outdoors but it does present a good starting point. I could show them Mary Anne Barkhouse's bronze wolves, which young people always enjoy. Instead, we stop by a small installation piece created with the help of various groups and students, entitled "the hyrdel." This is a small vine maze (slightly reminiscent of a labyrinth) created to walk through: a short winding pathway bordered on both sides by a crazily woven fence. It is about 24 inches high and just wide enough for one person. As they walk through in a line, I wonder what their response will be. The few ooohs and ahhs and "this is cools" are a bit of relief. Wild pedagogy gives me the confidence to proceed sometimes when it would be easier not to try to include the wild in an already packed agenda.

Trouble with the "Hyrdel"

It wasn't clear why I was being asked to take this work down. Although, it had started to look shabby. Somehow this outdoor eco-art work, the hyrdel, is part of my wild pedagogy. Eco-art put simply, is art that improves our relationship with nature and I have always felt wild pedagogy was a good fit with my interest in it. Wild pedagogy is in how the hyrdel was collaboratively hand-made; it is in the use of vine and sticks, both unwanted plants; it is in the challenge of keeping it; and it is in making it part of the Gallery tour. The hyrdel provided an opportunity for youth to consider a different creative process and a different aesthetic. The elements and principles of design are not all that applies here, and visitors experience creative work that can occur outdoors and made with our hands. It is a statement about art and art making. Wild pedagogy seems to embrace thinking and acting beyond the norms. If an art gallery is open to stretching one's thinking and perceptions then wild pedagogy is a welcome fit for an art educator.

As we head back towards the building, I have them look up the trunk of a tall spruce. I want them to observe a clay face up in the tree. By some miracle the face made with raw clay and never fired, has survived for over a year. It is the last of 37 sculpted faces placed up in the trees around the Gallery and like the hyrdel, it persists long past its expiration date. And I persist in pointing it out. I admire its survival. This work creates the sensation that the tree is looking at us, part of an installation called "conversation decay" by local artist Matt O'Reilly. What is the conversation that is in decay? Part of the answer is part of the reason I embrace a wild pedagogy. That conversing with the wild has been in decay, *is* in decay, but it also persists in the outdoors and with impermanent works that are made of natural materials. Being with these art works is part of acknowledging that conversation with the wild; therefore, the wild pedagogue in me allows for starting a gallery tour in this way. The teacher points to the face, the children look—and they are surprised by it, and enjoy being so.

We don't have much time and usually words are few, explanations bare, but I also don't mind letting them wonder. This is also what wild pedagogy has done for my practice. It is about giving an experience and saying less, about taking a chance on the response of the students to artworks or art making. Art works are objects of knowledge and parallel to nature can indeed be "co-teacher." After the tour I am out in the back lane collecting more vines, and since writing this, the "hyrdel" has been freshly woven and hopefully kept from being taken down. I know there are still sighs and a bit of eye rolling (from adults) but let them wonder too.

The Seeing of Rocks

It is part of the job of an arts educator to help students understand the materials they are using for their art projects, but I stumble when trying to explain rocks. The rock has often been used as a canvas, just a surface. When I found a rock left after a summer art camp that was smothered in acrylic paint I couldn't believe I actually scrubbed off the paint. I am reminded of a poem I found that describes a rock playing dead. Can we learn to see a world that is that alive? My friend scoffs at this, saying it is like going back to ancient times when the world was regarded as fully alive and it was rather frightening. But, in this world of teaching we can at least give the animals voice, the plants voice, and even the stones voice. I recall writing that we have to speak up for those that have no voice and in one way this is wrong—there is voice but it is us who have to listen. We can speak up but we must also listen. We listen with great attention to pings of our phones and act immediately on signals from our devices, so why is it strange that the world also has a kind of voice? Wild pedagogy seems to open up that door in its premise that nature might act as co-teacher. The wild pedagogy course and the floating colloquium on Wild Pedagogies helped me see the world as more alive.

These gatherings gave me confidence to keep telling my Christi Belcourt story. She is a Metis artist that had an exhibit at the Gallery. One of my favourite parts of working there is speaking with the exhibiting artists and I usually ask about what they would have me tell viewers as a message from them. One very large painting by Belcourt had many plants in it and I told Christi that children didn't seem to know the names of plants and showed little interest in knowing. I asked what I might say to them after they had taken time to look at the painting. She paused and then said, "Tell them…. to introduce themselves to a plant." Wild pedagogy would align quite nicely with this idea since it immediately addresses our relationship with nature. Art educators may be assured that many artists are potential strong allies in following a "wild pedagogy."

And Again

It happened again today—I chose to step outside to greet the high school group and mention that many artists are inspired by what they encounter in the outdoors and I guide them back down the sidewalk to look at the face on the tree. It feels natural and I feel more confident about doing these things because of wild pedagogy. I do the gallery tour and later I notice an artist has come to work on a piece that is already on the wall. It surprises me to see him working on the painting, although artists do last minute edits, even after their work has been installed in the Gallery. I mention this because when I asked the artist what his work was about he said tree spirits. I like this. Today two 19 year-old youth say these paintings are their favourite work. When I ask one why, he is silent then replies, "Because when I look at this work, I feel something is about to happen."

I love the fresh response of young viewers, and I am suddenly aware that, yes, the work does have an aliveness and now that he has said this ... "something is about to happen!" Wild pedagogy insists on allowing the spontaneity of such a response and the spontaneity of my own reaction to a student's words. Who is the teacher? How fun it is that I am nudged awake again. Nature as co-teacher, student as co-teacher, teacher as co-learner.

Reflecting on Wild Pedagogies: The Yukon River, an Inspiration for Practice

Victor Elderton is a career educator who has served as a teacher, and as a School Principal at North Vancouver School District's Outdoor School (re-branded as the Cheakamus Centre in 2013). He is currently a PhD student at Simon Fraser University. Victor participated in the 2014 *Wild Pedagogies: A Floating Colloquium* that took place on the Yukon River.

* * *

As a career educator, it is the Wild Pedagogies colloquium on the Yukon River I am drawn back to. I'm quick to discuss it with friends and colleagues. On my experiential learning playlist, it's in the top five. To-date the experience informs and inspires me, personally and professionally. My thoughts about it swirl like eddies. But, what is it about the experience that continues to wash over me? What is it that continues to create meaning and purpose in my learning?

No doubt it is the wildness—and the unpredictability of it—that makes wild pedagogies engaging, accessible, and meaningful for me. However, such meaning is not possible without intentionally setting the stage. As an educator and administrator I seek to incorporate ideas from exceptional instruction and experiences into my practice and theory. I argue that incorporating what has been experienced in one context and then making it useful in my own context is an effective way to improve practice. I admire instruction that is intentional and effective—there is an art to it, and a potential mastery. I strive for these artful goals, and wild pedagogies, as ideas and practices, helps me by exhibiting strong characteristics of intentional design. Attributes of wild pedagogies colloquia that stand out for me are their capacity for:

- Convening gatherings of like-minded, self-selected learners. By accomplishing this, the enthusiastic opportunity for shared learning is possible.
- Providing these groups with a topic to collaboratively investigate. With this, integration of multiple perspectives and appreciation is likely.
- Co-creating an itinerary with specific, and emergent, opportunities to discuss, collaborate, and exchange impressions and ideas. In this way intrinsic motivation is owned by the learner, what is being learned becomes more relevant.
- Setting the group on a physical journey through an inspiring place. Here, the wildness of the place becomes a co-teacher and the experiences give it a voice—makes it tangible.
- Utilizing immersive experiences, and human and other-than-human perceptions, to open opportunities for wider and deep understanding.

There is a high level of integrity in this approach to learning and intrinsic motivation.[1] Wild pedagogies can reinforce this instructional design in daily practice.

On a typical day, my classroom starts in a circle discussing how students are motivated. Each student has an opportunity to talk about what inspires them and what questions they want to ask. We also discuss how the day will unfold and develop a rationale for the order. From that starting point, both inside and outside locations become our classroom. We discuss what each of us will need to bring to our learning to make it meaningful and intentional. My skill as the maestro on this daily journey is to bring out the flavour of emergent opportunities, enabling them to flourish in daily learning. At the end of the day, we gather again in a circle that allows each student, me, and other staff members to wrap up our learning for the day. This is how wild pedagogy and my practice are linked. It's when I incorporate newly arising ideas into my teaching and struggle with them in application that truer instruction becomes apparent and useful.

Another key learning from Wild Pedagogies is its emphasis on being a hands-on endeavour—improving by doing. I don't think that this is just a function of Gladwell's ten thousand hour rule.[2] There is something deep and transformative in physical learning. For centuries philosophers and educators have identified unique learning and understanding that is intrinsic in doing. Confucius identifies charioteering and archery as two of the six disciplines essential to all-around development, Plato highlights gymnastics. Saint Francis believed that starting with the necessary makes the impossible—possible, Booker T. Washington developed his Tuskegee institute based on deeply learning a skill, Dewey and Montessori use direct experience as cornerstones of their educational practice and theory.

Wild Pedagogies on the Yukon River was instrumental in the continued development of my sense of education. It helped to illuminate essential ingredients to learning and teaching: motivation, inspiration, context, and dialogue. Wild Pedagogies exceeds the structures we have traditionally built for education, and that Montaigne argues against when he reflects on an education that: "cudgelled my brains in the study of Aristotle."[3]

Wild Pedagogies on the Yukon River provided me with an opportunity to investigate my learning and understanding while physically immersing myself in a place-based context. This context demanded I be consciously and contemplatively present, simultaneously. While paddling, or in the midst of teaching, there are requirements to watch the currents and navigate a course. In a teaching setting the individual needs of each student demand attention, as does the weaving together of these needs to become the course. The Yukon experience occurred while passing through, and being in, a majestic landscape, with glimpses of other presences such as mother moose ferrying her new calf across the river, or sheep traversing a precipitous trail high above our river course. Similarly, in school a child may notice a bird building a nest in the tree outside class, and then talk about how hard it must be to build a house and create a home. It's in these teaching situations where motivation, inspiration, context and dialogue are holistic. It is in these structures for learning that I find the most meaning.

I sought active learning within intentional instruction. I wanted to explore learning with rich opportunities for self-motivation, inspiration, context, and dialogue. I received all this from *Wild Pedagogies: A Floating Colloquium* and it's why I continually draw from its deep currents to inform my practice.

Project Wolf: Re-wilding Head, Heart, and Hands

Joyce Gilbert works for the Scottish Non-Governmental organization *Trees for Life*. It is a small conservation charity that seeks to restore the ancient Caledonian forest in the Highlands of Scotland. She has enormous experience across a broad range of educational settings, including formal schooling and with Non-Governmental Organisations. With her knowledge of local cultural history and her studies in the Gaelic language, she was a key resource person during the *Wild Pedagogies: A Sailing Colloquium* in May 2017. She begins her story by making connections between her present work with *Project Wolf* and wild pedagogies. She then draws on the touchstones put forward to identify possible future directions for continued development of this project.

* * *

In 2016 and 2017 *Trees for Life* embarked on an unusual initiative to aid the natural regeneration of wild forest in unfenced areas of their Dundreggan Conservation Estate. Called *Project Wolf*, dedicated groups of 3 volunteers came together for a month at a time in early spring when young trees are most vulnerable to deer damage. Sleeping during the day and emerging at dusk, the volunteers spent the night mimicking aspects of wolf behaviour to keep deer on the move and prevent grazing of the seedlings. Scientific methods were used to determine the impact of the "wolf interruption" and after two seasons, this innovative approach to natural regeneration appeared to be making a significant difference.

However, what hadn't been anticipated was the profound effect the experience would have on the volunteers. Within a few days, the first "Pack of Humandwolves" (as they came to be known), had chosen and named campsites, resting spots and significant landmarks and started to experience nature in a new way. Time was suspended in this strange night world and senses were heightened. One of the early *Project Wolf* volunteers described her experiences in a way that resonates with the wild pedagogies idea:

> As a wolf, you must do more than rise to the challenge you have been set. You must look at the land and imagine what it would have been like. Re-wilding happens in heart and the mind as well as on the ground. Sitting at the top of *Binnilidh Mhor*, I see a land cut into fragments. But I also see the woodland, the wolf pack's territory, clinging to the hills below us, and think how wonderful it will be in 30 years' time. And words cannot do justice to how it feels to think I had a hand in that.[4]

With the experiences in the landscape came a deepening awareness of what it is to dwell in a place and the ability to recognise the complex relationships between trees, deer, predators, prey, but from the perspective of humans that are not merely observers of nature but playing an active part in it. Another volunteer described her emerging awareness as follows:

> On my nights spent as a wolf I really felt like I'd become part of the woodland. I was just another creature roaming around becoming familiar with its nooks and crannies. I had my favourite spots to sit and watch from as well as my favourite trees and sleeping places. I already knew I enjoyed spending time in wild places.... Being a wolf showed me that what's important is the feeling of connectedness that comes with living closely to the land. It also made me realise that this feeling of connection has been missing from my city life in recent years.[5]

As a development of the project, *Trees for Life* commissioned the design and creation of a "wolf den" by a young artist called Richard Bracken who worked in collaboration with the poet and artist Alec Finlay. Natural materials for constructing the den were sourced from the immediate environment. The idea was that this structure could be used by Humandwolves for shelter, but it would also would invite people to view nature, wildness, and wolves from different perspectives. Richard and Alec developed a number of concepts, ranging from very light shelters to more substantial dens. They researched traditional shelters in Gaelic culture, such as shieling huts and *phouple* (tents used in hunting). In addition, they surveyed Gaelic place-names and garnered information on species, to create an "ecopoetic" mapping of the region. The site chosen was an elevated flat "table" covered with heather, and with juniper and birch around its sides. This provided a well-drained platform for a structure on a fairly prominent landscape feature that looked directly across the glen to *Creag a' Mhadaidh*—meaning the crag of the wolf, in Gaelic.

Alec Findlay described in more detail how paying careful attention to Gaelic place names could, in fact, give voice to the eco-social history of the place:

> During the research phase, Richard and I considered working from existing natural forms, such as hill-shaped walls that remain permanently in place, which a tent-like covering can be attached to when needed—a memory of when folk would carry sails uphill to shielings, where they became a roof for the walls of a hut. We also developed the concept of tent-shaped hills—

these could be based on local names such as *Creag a' Mhadaidh*, wolf crag. The place-name research included a survey of wolf-related names such as *Ceap Mad*, root-bog of the wolf. We defined this approach as "eco-poetic" mapping.[6]

The final design was informed by talking to individuals who had undertaken the role of the Humandwolf, asking what they would want for shelter, how they had bonded, and how their experiences had altered perceptions. The unusual nature of the questionnaire devised by Alec provided some fascinating insights with respect to "wolfing:"

> It became clear that the performative and imaginative aspect of the Humandwolves was highly relevant to concepts of re-wilding and human relationships to nature, indeed, we consider it to be some of the most important field research being undertaken in the Highlands.[7]

In retrospect, the similarities between the 1500 year old "beehive cells" (described by James Hunter in his book *On the Other Side of Sorrow*, and visited by the *Wild Pedagogies* group in May 2017) and the Wolf Den at Dundreggan are striking—in terms of form, function, and setting. However, the den is only one part of *Project Wolf*. Alec Finlay has noted that something deeper was revealed by the Humandwolves in their need "to dwell" and furthermore, this has the power to change how people respond to place—in Scotland at least:

> Place-names insist on people's right to have access and, if we accept that for now that battle has been won, the issue that must follow is dwelling. When the meanings of names are translated it can encourage communities to go beyond the pleasures of walking and shooting, and take on the responsibilities of stewardship. Summer-towns cared for many remote places—whether as tenants or squatters—and recovering their names is one way to suggest this could happen again—that you too could regenerate a patch of Glenmoriston, Glen Feardar, or Glen Affric. Scotland's become stuck in an argument about land ownership when what's really needed is for the right to care for a place to be added to the right to roam over a place.[8]

As a teacher who was part of the *Wild Pedagogies Sailing Colloquium*, I was curious to revisit the questions we had been asked at the beginning of our journey—what else, for example, might *Project Wolf* offer to educators who are interested in re-wilding the curriculum with nature as co-teacher? Based at Dundreggan, I had already started to work with local primary and secondary schools and community groups. At the most practical level, could I use the questions and wild pedagogies touchstones to help with curriculum design? Working with Alec and Richard for a relatively short time, the following preliminary ideas emerged. Based from *Trees for Life* we could:

- Produce a "wolf map" detailing natural howffs (rendezvous locations), good juniper bushes to shelter in, path networks, and other landmark features;
- Rethink a *Tainchell* (Gaelic for a traditional deer drive). This could entail a collective "drive" survey and communal walk to a single point on the hill. As an ecologically-minded activity, it could focus on progress in growth and regeneration;
- Organize special "Wolf Walks" which follow some of the tracks used by deer at Dundreggan and involve discovering nocturnal animals, insects, and birds; viewing the night sky; and visiting the Wolf Den story telling;
- Work with locals and visitors to create a sound installation that will capture elements of Project Wolf (e.g. woodland soundscapes at night, conversations about wolves and other 'lost' wildlife stories);
- Work with schools and communities to create poetic and story pieces (including traditional Gaelic nature-praise poems) that will link with the "ecological memory" mapping, wolves, and deer. In the process, new place-names and new stories will be created; and,
- Introduce "wolfing" as a new activity at Outward Bound Schools and Outdoor Education Centres.

Project Wolf started with the simple aim to reduce deer browsing young trees at a critical time. However, it was given the freedom to respond to the collective imagination of ecologists, educators, cultural historians, native Gaelic speakers, and place-aware artists, and poets—it "wiggled in its own wildness" and in the process it was transformed.

Modern Education in Japan

Yuko Oguri works in Higher Education at Kagoshima University in Japan. She specializes in community-based education and traditional knowledge. Her concern is the extent to which modern education in Japan is controlled in ways that support a nationalised agenda and, at the same time, alienate people from their lands and traditions. Yuko participated in Yukon's *Wild Pedagogies: A Floating Colloquium* in 2014. For her, wild pedagogies is a term that helps to reframe a vision for education that reconciles "pre-modern education" and "modern education."

* * *

In Japan the concept of modern education was introduced after the Meiji Restoration in 1868. This was a time of social and political change and the beginning of the collapse of shogunate government. It was replaced by an Emperor system and Japanese unification. It also marked a shift from a feudal to a capitalist society. As part of this change, a new national schooling system was initiated in 1872. The purpose of this new educational system was to "control" people such that they would become members of a single nation. This was by design meant to deny, in the raising of children, the diversity of culture that belonged within each community and the unique natural environment in which each community was embedded. The outcome of such a well-controlled educational system was an "equally" developed nation with a "hundred million middle-class" citizens. There remained, however, many contradictions between old and new systems, including in peoples minds and values.

The book, *Chiho Shoumetu*, roughly translated as *Cities at Risk of Disappearing*, by the former Minister for Public Management,[9] created quite a sensation when it warned that 896 Japanese municipalities, mostly in rural areas, out of a total of 1718, will disappear due to decreasing populations. This "official opinion" stirred up a sense of crisis among communities. However, the government's failures in national land use planning, and their development policies that accelerated the population decline, are seldom mentioned. Furthermore, there is no reflection on the merits and faults of the schooling system, and the role that it plays in enabling these trends.

As a Japanese educator living close to cities at risk of disappearing, I see a somewhat different trend and possibility within my community. I work with local leaders and officials to help foster a learning environment that encourages community members to share their worries and desires, and to support their collaborations aimed at developing livelihoods. I am always astonished by the strong ties that exist amongst these people, and with their land. In spite of the widespread loss of "places for self-formation,"[10] meaningful educational spaces do still exist in most cities. In the communities where I work, there still are places where experiential learning can take place, and there are places for activities that can encourage positive relationships between children and nature—and between other beings and things.

For example, in rural areas of Japan there is a strong will, and much effort given, to pass along traditional cultural activities and events that are Indigenous to specific places. The desire to pass along these year-round activities is especially strong in more remote islands like Amami. This cultural learning is a remnant of Indigenous living that once existed before Japan became modernized. Sometimes it is characterized by religious animism and the blessing of and caring for nature throughout the year. Both informal and non-formal education is engaged. Village Elders pass down their knowledge to younger generations. They share collective stories and demonstrate how seasonal events are prepared, performed, and tidied up, thus restoring the village's relations with its nature and people. These processes are usually overlooked as being "educational" in a schooling system that has disconnected people from the land for over 150 years.

The movement I am describing is not just going back into the past. Rather, it is a process of forming a new future vision of "education" that seeks to reconcile aspects of "pre-modern education" with modern education. In Japan, we don't have a correct word to express this new vision of "education," but I think the idea of wild pedagogy perfectly names it.

What is appealing about wild pedagogy is that it allows us to more deeply seek that alternative education which still exists. And wild pedagogies can help to describe the educational value and meaning embedded in the activities that are actively sustaining Indigenous culture. Using this term will help us to share and appreciate a search for a new vision of education within Japan, and perhaps beyond.

Early Childhood Education

Andrea Welz describes herself first as an educator and parent. She is also a faculty member in the Early Childhood Educator Program at Sault College in Canada. Andrea has brought her experiences with wild pedagogies to bear in her teaching, in policy development, and in establishing a preschool programme offered in a natural setting. She first learned about wild pedagogies during a graduate course at Lakehead University in 2012. She also participated in 2014 *Wild Pedagogies: A Floating Colloquium*, in the Yukon, and the 2017 *Wild Pedagogies: Sailing Colloquium*, in Scotland.

* * *

I love the intense conversations that arise whenever wild pedagogues get together. The concepts discussed stretch my perspectives such that new ideas and ways of thinking emerge. But I must admit, throughout this process I wonder what these concepts would look like in the world of early childhood education. Putting these insights into practice is exciting, both in my role as an educator working with children and as a college faculty member.

I think the recent pedagogical shift in early childhood education makes it receptive to wild pedagogy. It asks educators to become more reflective in their daily practice, specifically about their image of the child and the environment as teacher. And the call for building equitable relationships, where children are co-constructors of their learning, parallels some of the wild pedagogy offerings. With this in mind, I am working with a licensed childcare centre to develop a nature-based early childhood program in our community that encompasses wild pedagogy.

The plan is to offer a half-day preschool program in a natural setting. It will be play-based with lots of time for children and adults to fully engage in this natural environment. Along with this focus, the program will function as a parent cooperative; this means that two to three parents will be scheduled to work with the early childhood educator each day. Families will be involved in the evolution of the programming that will include using the wild pedagogies touchstones to guide our practice.

Elders have also been invited to participate and their involvement will be an integral piece of this project. I envision a program where children and adults might be found, either on their own or with groups, playing in puddles, gently touching moss, watching ants as they carry material back to their colony, building little shelters, working on projects like basket-making, listening to stories, or tending a fire. Adults, with the children, will narrate the stories about the learning that is taking place. Adults will share their reflections about how the touchstone questions shaped the way they engaged with the children and the more-than-human world, including the challenges and queries that arose. I think that these questions can also guide routine activities such as the selection of the stories and books that will be shared with children. It will be a process that will take time.

I found that wild pedagogies also helped me to shape the programme's policies and procedures. In Ontario (and many other provinces in Canada) an adherence to legislative requirements, and the policies and procedures that arise from them, is a critical piece of a programme. As dry as this topic may seem, these policies and procedures do frame the programming offered. One key policy arose in our planning because of concerns about black bears. This is a fear that often dominates discussions about, and at times deters, nature-based programming. The policy could have been very human-centric with a tone that sees the rest of the natural world as lesser-than-humans. However, using the wild pedagogies touchstones as a guide, a policy was created called "Sharing Spaces." It acknowledges that natural settings are spaces that are shared by many. Carefully selected wording emphasizes that all-of-the-natural world is valued. It is important to note that the policy does not negate the safety concerns regarding children; several strategies have been developed to manage risks. But it doesn't stop there; the policy also includes strategies to minimize impacts on the rest of nature (e.g. frequently changing locations to avoid compaction of the soil). It is one tiny step forward in shaping a perspective that builds equitable relationships with all entities on this planet.

Integrating wild pedagogies into the college early childhood education program has been a bit more challenging. At first I wondered how teacher-directed experiences, designed to meet course-learning outcomes, might include wild pedagogies. I scheduled some classes in a forest area close to the college and, although planned experiences were still tied to learning outcomes, I observed, or students told me, how these experiences affected them in unexpected ways. For them, nature did, indeed, become a co-teacher. Inviting the Elders-in-residence to share Indigenous perspectives about nature and child development has also been more meaningful when we have met in the forest. They offer students an opportunity to learn about other ways of knowing and understanding the world around them. Doing this with a direct connection to the land added another dimension.

Another activity involved inviting college students to select a space they were drawn to so they could sit quietly on their own for thirty minutes: this was an experience that was initially disconcerting for some. A more task-oriented experience required students to look for ways that math and science concepts could be fostered in indoor, playground, and natural settings. Over the past few years students have overwhelmingly described these tasks as being more enjoyable in the forest where uncovering math and science concepts became intriguing and engaging. This is illustrated by a group who found little bead-like nodules in a bank along a stream. Time seemed to slow down as the students touched the nodules and discovered that they were connected to a root system. They marvelled at this intricate system and then began to question and hypothesize. Watching them in that setting, enveloped by the rustling of the leaves and the gentle murmur of the stream, I had a sense that there was more happening than just the meeting of a science-focussed learning outcome. In some cases this learning was something the students themselves might not be able to explain.

Sharing these ideas and subsequently dialoguing about them with different groups of educators and parents has been beneficial. It has led to both expanding my understanding of wild pedagogy and building a supportive community as we shift our ways of being in the world.

On Nature as Co-Researcher

Sean Blenkinsop is a philosopher of education and a Professor at Simon Fraser University. He has a long background in experiential, environmental, and outdoor educations. Sean has participated in all of the Wild Pedagogies colloquia to date.

* * *

If we take the touchstones for wild pedagogies seriously, then what implications might they have not only for educators but also for researchers? It is a question that has been bubbling for a while and in some ways wild pedagogies in Scotland was a tentative response. I say tentative because it is clear to me that I still have a long way to go in terms of listening to and understanding the voices and research agendas of more-than-human others, and in representing the results in genuine, just, and nuanced ways. For wild pedagogies, getting outside the cultural norms of public education is important. But research has norms too. In fact much of what is considered research in universities appears grounded in the same norms that created public education and pushed us into the Anthropocene. So, it follows that there might be something called "wild researches" and wild researchers. They, too, should be asked to push against troublesome cultural norms, become activists, build rich communities, and engage with the natural world in different ways. But we might also be asked to shift ourselves, our questions, and our methods away from the centre of contemporary research. In keeping with wild pedagogies, wild researches could consider research subjects more as partners not objects and hence, come to practice, present, and understand research differently.

My sense is that this book is a tiny step in that direction. There is an implicit comment on the mainstream concept of knowledge and the ownership of ideas. There is also an attempt to represent and recognize more voices than just those of the humans. There is a critique of conferences, particularly environmental ones, held in hermetically sealed hotels where the wild has little to no access. There is also a desire to move to a more dialogical form of sharing and away from the 15-minute presentation, with five minutes for questions, format. And finally, there is recognition that places afford differing possibilities to human theorizing (e.g. we chose Scotland carefully or note the on-going metaphoric presence of the sea) and we have taken a crack at speaking in a multi-vocality that reaches and represents a wider constituency. Have we gotten it right, nope … are we on the right track … well … we hope this contributes to growing challenges to research norms.

Questions I am beginning to ask myself as a wild researcher:

- How does the natural world ask and answer its own questions? What are its accepted methodologies?
- How do/did/might I engage with other-than-humans and represent them in my/our work?
- Where is the natural world positioned in my research? To what and whose end?
- Have I tried to represent my findings in a way that does justice to the contributions of others?
- In what ways did I enter and engage with research locations? Might there be room in community-based research methods to include more-than-human communities as full members?
- Where is my research maintaining anthropocentric forms?
- What are the implications of this research for the natural world?
- How do I deal with what seems to be the researcher's paradox—balancing being present and listening to the other-than-human against disappearing into my own thinking?

But, really I am just a beginner here … what do you think?

Going Forward

In closing this book, we want to stress that the work begun here is not complete. Indeed, we now wonder about a corollary to wild pedagogies, that is "wild learnings." Throughout we have argued that wild pedagogues must be co-learners and the examples in this chapter illustrate the point. But what could this mean if we thought more broadly about learners through the lens of wildness proposed in this book? Ultimately this is a question to take forward. However, while writing these final pages a new book presented itself, quite serendipitously, that might point us in a possible direction. In it, Kate Harris has openly told a story about re-negotiating her owning learning practice:

> I was too good at school, in every doomed sense. After being on an achievement bender most of my life, the prospect of withdrawal, of doing anything without approval, or even better yet acclamation, kept me obediently between the lines I couldn't even recognize as lines. Isn't that the final, most forceful triumph of borders? The way they make us accept as real and substantial what we can't actually see?[11]

She speaks about how being good at school earned her a scholarship to Oxford. However, the learning environment she found on arrival was liberating: "once I got there I almost learned not to care about [school], or rather to care for the right reasons: not as a means to [an] end, or success sanctioned by others, but as an opportunity to think, dream, stray out of bounds."[12]

Harris also found professors who "encouraged digression, which is, after all, just a sideways method for stumbling on connection. Such as the connection between the philosophy of science and poetry."[13] And for her, the best part about studying the history of science was that suddenly she had, "to do for homework what I normally did for fun: read expedition journals, such as Charles Darwin's from his voyage on the *Beagle*."[14]

In reflecting upon her experiences with traditional education Harris asks, "What does it mean when you build your own walls? You have no one to blame but yourself for inhabiting them."[15] While Oxford is a renowned university, the question is not exclusive. In it, we have beginnings of a nascent touchstone that asks wild learners:

- What opportunities have I taken to think, dream, stray out of traditional educational boundaries?
- Have I crossed any borders that have re-excited my learning, and even made education fun again?

Our immediate reaction to this last story is that it encourages us, and readers to consider all of the touchstone questions as if being posited not just for teachers but for learners as well. It invites us to revisit the touchstones anew. This story can also be seen as another invitation to take this book as an opportunity to think, dream, stray outside of traditional educational boundaries. That would be a splendid expressive outcome.

In many ways writing this book has been an experiment—in ideas, in practices, and in collaborative scholarship. We have gathered together, in these pages, the scent of potential change. We hope, though, that there is enough here for readers to get started—or to keep going if they are already on their own journey of transformation. And, as David Orr often says, "hope is a verb with its sleeves rolled up."[16] Our journey isn't a passive one.

One thing that we are fairly sure about is that we will not figure out what is to be done, or how to do it, in advance. This, ultimately, isn't an abstract project; it is real, and with real implications. And, we will learn most from doing—experimenting, trying things out, and making new colleagues, human and more-than-human. Again we hope that *Wild Pedagogies* will give readers permission to think anew, to try new practices, to get outside, to question deeply help beliefs, and to disrupt long-held traditions. So, wild pedagogies do not represent concrete frameworks or destinations; rather, they are agents of discovery. Good luck in your explorations. And thank you.

Acknowledgements *Crex crex* is the taxonomical name given to the Corncrake. We have chosen this bird to represent our collective because it was an important collaborator in this project and because its onomatopoeic name beautifully mirrors its call—a raspy crex crex. For some reason, it chooses to fly over England and breeds in Scotland and Ireland. Presumably this is due to loss of habitat in modern England, but perhaps these birds sense some epicenter of empire there? Who is to know?

Notes

1. Mihaly Csikszentmihalyi, *Beyond Boredom and Anxiety* (San Francisco: Jossey-Bass, 2000).
2. Malcolm Gladwell, *Outliers: The Story of Success* (New York: Little, Brown and Co., 2008).

3. William C. Hazlitt, ed., *Essays of Michel de Montaigne*. Translated by Charles Cotton (London: London Reeves and Turner, 1877).
4. Millie Barratt, Personal communication, 2016.
5. Liv Glatt, Personal communication, 2017.
6. Alec Finlay, Personal communication, 2017.
7. Alec Finlay, Personal communication, 2017.
8. Alec Finlay, *Of Wolves and Men*, forthcoming.
9. Masuda Hiroya, ed., *Chiho Shoumetu* (Tokyo: Chūōkōron-shinsha Inc., 2014).
10. Takahashi Masaru, *Self-Formation Space of Children* (Tokyo: Kawashima Shoten, 1992).
11. Kate Harris, *Lands of Lost Borders: Out of Bounds on the Silk Road* (Toronto: Alfred A. Knopf Canada, 2018): 63.
12. Ibid., 52.
13. Ibid., 47.
14. Ibid., 48.
15. Ibid., 65.
16. See for example: David Orr, *Hope is an Imperative* (Washington: Island Press, 2011).

References

Csikszentmihalyi, M. *Beyond Boredom and Anxiety*. San Francisco: Jossey-Bass, 2000.
Gladwell, Malcolm. *Outliers: The Story of Success*. New York: Little, Brown and Co., 2008.
Harris, Kate. *Lands of Lost Borders: Out of Bounds on the Silk Road*. Toronto: Alfred A. Knopf Canada, 2018.
Hazlitt, W. C., ed. *Essays of Michel de Montaigne*. Translated by C. Cotton. London: London Reeves and Turner, 1877.
Hiroya, M., ed. *Chiho Shoumetu*. Tokyo: Chūōkōron-shinsha Inc., 2014.
Masaru, T. *Self-Formation Space of Children*. Tokyo: Kawashima Shoten, 1992.
Orr, D. *Hope is an Imperative*. Washington: Island Press, 2011.

Index[1]

A

Abram, David, 36, 38, 47n21
Alliances, building, 101–106
Amami, 123
American Educational Research Association, 4
Amy, 68–70
Angeconeb, Ahmoo, 111
Animals, human relationship with, 32
Animism, 123
Antarctica, 39
Anthropocene epoch, 2, 19, 52
Ardtornish bay, 82
Aristotle, 116
Arts, 69, 110–113, 115
Atlantic Ocean, 42

B

Barkhouse, Mary Anne, 111
Battiste, Marie, 33, 47n20
Bauman, Zygmunt, 5, 6
Bears, black, 125
Being, viii–x, xii, 2–4, 6, 7, 9, 11, 24, 26, 27, 29, 30, 34, 36, 39–41, 44, 52, 53, 56, 59, 67, 70, 71, 73, 78, 80, 82, 83, 86, 87, 91, 92, 94–97, 99, 100, 102–105, 110, 112, 113, 115–117, 119, 120, 123, 126, 128, 129
Belcourt, Christi, 113
Binnilidh Mhor, 118
Bissell, Norman, 106
Blenkinsop, Sean, 61n5, 75n1, 107n3, 107n4, 127

---

[1] Note: Page numbers followed by 'n' refer to notes.

Body language, 89
Bokova, Irina, 5, 6
Bracken, Richard, 119
Brendan, 16
Bringhurst, Robert, viii
Buber, Martin, 99

Canada, 18, 31, 39, 111, 124, 125
Canadian Network for Environmental Education and Communication, xi
Capitalocene, 58
Caribou, 32, 43, 111
Ceap Mad, 120
Change
   climate, 2, 52, 67, 100, 104
   cultural, 13, 97, 99
   earth, 3
   educational, xxix, 5, 18
Cheakamus Centre, 114
Chiho Shoumetu, 122
Chthulucene, 58
Cities at Risk of Disappearing, 122
Civilization, 35
Climate change, 2, 52, 67, 100, 104
Cnoc Buidhe (Yellow Hill), 27–29
Colonialism, 13, 24, 26, 29, 31, 44, 46n19, 58, 88, 91, 98, 102
Columba, Saint, 16, 97
Community
   human, 71, 101–106
   oppressed, 3
   wild, 102, 104
Companionship, 11
Complexity, 31, 35, 70, 84–88, 90
Confucius, 116

Consciousness, 33, 44
   Eurocentric, 33
"Conversation decay", 112
Co-option, xii, 36, 58, 67
Corncrake, 15, 16, 90
Creag a' Mhadaidh, 119, 120
Cronon, William, 25, 26, 30, 31, 34–36, 40, 43
Cuckoo, 15, 16, 90
Cultural change, 13, 97, 99
Curriculum, x, 18, 33, 65–67, 69, 83, 93, 99, 101, 102, 121

Darling, Frank Fraser, 13
Deep time, *see* Time
Democracy, 102, 103
Devonian epoch, 97
Dewey, John, 116
Discipline, 92, 94, 95, 116
Disenfranchisement, 29
Dolphins, 104, 105
Doug, 57, 72
Dualism, 33, 35, 37, 40, 47n21
Dundreggan Conservation Estate, 118

Eagle, 82, 83, 95
Early Childhood Educator Program, Canada, 124, 126
Earth
   agency, 6, 55
   change, 3, 52
   geostory, 53–56, 58, 59, 78, 94, 97
   human relationship with, 6, 24

'quaking', 53
state of, 52
Easdale slates, 101
Eco-art, *see* Arts
"Ecology without nature", 36
Eco-poetic mapping, 120
Eco-social learning, xii, 9
Education, 118
   change, ix, 5, 18, 81
   control and, viii, 2, 20, 86, 122
   domestication of, x, 2
   early childhood, 73, 111, 124–126
   etymological origins, 69
   models of, viii, 4, 84
   nature-based, xi
   reform of institutions, 24
   re-wilding (*see* Wild pedagogies)
   role and value of, 20, 69, 124
   testing, x
   wilderness and, 19
   See also Environmental, education; Outdoor education; Sustainability education
Education for People and Planet: Creating Sustainable Futures for All, 5
Eileach an Naoimh, 15
Eisner, Elliot, 65
Elderton, Victor, 114
Emotions, 70, 71, 74, 78, 91, 105
Environmental
   crisis, 58
   education, 53, 66, 67
   etiquette, 4
Experiential learning, 5, 115, 123
"Expressive outcomes", 65, 130
Extinctions, 3, 52, 67

F

Fingal's Cave, 17, 84
Finlay, Alec, 119, 120
First Nations peoples, 31
Forest School Movement, xi
Francis, Saint, 116
Freedom, 2, 27, 29–32, 40, 42, 43, 60, 66, 102, 121
Fulmar petrels, 57–60

G

Gaelic poetry and culture, 15, 29, 32, 71
Garvellachs, 9, 15–17, 92, 93
Geology, 2, 3, 17, 19, 40, 41, 52, 58, 59, 78, 79, 91, 98
Gilbert, Joyce, 117
Gladwell, Malcolm, 116
Glen Affric, 120
Glen Feardar, 120
Glenmoriston, 120
Global warming, 55
   See also Climate change
Gough, Noel, 84
Greenwood, David, 38
Griffiths, Jay, xii
Gulls, 57, 82, 83, 87, 95
Gutiérrez, Kris, 4, 5

H

Habits, 33, 83, 90, 91, 93, 94, 96, 97
   changing, 93, 94
Haraway, Donna, 58, 59
Harris, Kate, 129
Hebrides, Inner, 15, 70, 78, 103
Heuristics, 19

Highland Clearances, 29
Highlands, Scotland, 29, 30, 117
Holocene epoch, viii, 19
Humandwolves, 119, 120
Humans
    centredness, 8, 11, 24, 34, 36, 37, 40, 58, 90, 91
    control, 6, 34, 56
    nature, relationship with, 32, 112, 113, 120, 123
Hunter, James, 13, 15, 29, 30, 32, 33, 120
"Hyrdel, the", 111–113

Imagination, 5, 19, 25, 44, 60, 68, 83, 86, 104, 121
Imperialism, 13, 30
Indigenous peoples, viii, 25, 29–31
Industrialized societies, 24
Inner Hebrides, 15, 70, 78, 103
Intervention, 4, 5, 98
Intuition, viii, 94, 95, 97
Iona, 14–16, 27, 97, 104

Japan
    modern education in, 122
    population decline, 122–123
Johnson, Chief Joe, 31, 36
Justice, 31, 64, 102, 104, 118, 128

Kagoshima University, Japan, 122
Kassi, Norma, 31, 32, 43
Kluane national park reserve, Canada, 31
Knowledge, 3, 9, 43, 46n19, 83, 84, 86, 88, 111, 112, 117, 122, 123, 127

Lady of Avenel, The, xii, 9, 13, 14, 17, 93
Lagandorain, 27
Lakehead University, Canada, xi, 111, 124
Land management practices, 13
Landscape
    learning in, x
    walking in, ix
Language, 4–6, 16, 19, 29, 31, 33, 35–37, 39, 44, 56, 78, 91, 103, 117
Last Child in the Woods, xi
Latour, Bruno, 52, 54–56, 58
Learning
    content-based, 69
    eco-social, xii, 9
    experiential, 5, 115, 123
    first order, 69
    ingredients of, 116
    re-learning, 71
    transformative, 64, 116
    See also Education
Listening, x, 8, 11, 12, 16, 32, 37–39, 44, 56, 72, 73, 80, 81, 87, 95, 113, 125, 127, 128
Lorde, Audre, ix
Louv, Richard, xi
Luing, Isle of, 12, 17

M

McKibben, Bill, ix
MacLean, John, 27
Marginalization, 3, 24, 29
Meditation, 74, 95
Meiji Restoration, Japan (1868), 122
Memmi, Albert, 90, 107n4
Monastic communities, 15, 16
Montaigne, Michel de, 116
Montessori, Dr Maria, 116
"More-than-human", x, xii, 2–4, 6, 7, 9, 11, 12, 24, 34, 36–40, 44, 56, 60, 74, 80, 81, 95, 100, 102, 103, 105, 125, 127, 128, 130
"More-than-plankton world", 36, 37
Mull, Isle of, 27, 91, 92, 94

N

Næss, Arne, 48n39, 89, 94, 107n2
National parks, 30, 31
Nature, 7
  concept of, 24
  as co-researcher, 7, 8, 127–128
  as co-teacher, 7, 8, 11, 64, 68, 74, 79–83, 112–114, 121
  critiques of, 25
  education and, xi, 24, 69
  gurus, 89
  human relationship with, 36, 90, 120, 123
  term, 35, 40, 46n19
  voice, *see* Voices
Neoliberalism, xii, 66, 67, 69
North Africa, 90
Northern Protected Areas and Wilderness conference (1993), 31

North Vancouver School District Outdoor School, 114
Norway, vii, viii
Not-knowing, 72–74

O

Oguri, Yuko, 122
Ontario, 125
On the Other Side of Sorrow, 13, 29, 120
Open mind, 8, 72
O'Reilly, Matt, 112
Orr, David, viii, ix, 130
Otter, 95
Outcomes, 5, 6, 65, 67, 68, 86, 101, 122, 126
  defined, 87
Outdoor education, xi, 67, 127
Outward Bound Schools, 121

P

Paleontological epochs, 52
"Paths Beyond Human-Centredness", 36
Pathways, xi
Pedagogies, viii, 4, 5, 18, 24, 44, 65–67, 69, 95, 99, 100, 104
  walking as, ix
  See also Wild pedagogies
Place
  names, 71, 119–121
  relationships with, 3, 7, 8, 97
Plato, 116
Play, vii, ix, 5–7, 11, 16, 30, 34, 55, 68, 72, 83, 87, 90, 94, 97, 98, 100, 113, 118, 123, 125

Pleistocene epoch, 79
Plumwood, Val, 36, 40
Political activism, 99
Power structures, 100
Practice, viii, x, xii, 4–6, 12, 13, 20, 25, 29, 33, 39, 44, 60, 66–69, 78–106, 110, 112, 114–117, 124, 125, 127, 129, 130
   See also Wild pedagogies, practice
Precambrian epoch, 88
Problem solving, ix, 59, 89
Profeit-Leblanc, Louise, 8, 9
Project Wolf, 117–122
Puffins, 17, 41–44, 68, 85

R

Reason, 32, 43, 44, 59, 60, 72, 88, 94, 102, 110, 112, 129
Rebel teachers, 64
"Relative wild", 36
Re-membering, 70–71
Re-negotiating, 4, 6, 24, 25, 27, 59, 84, 110, 129
Research, x, xiiin6, 65, 66, 79, 111, 119, 120, 127, 128
Resource extraction, 102
Responsibility, 4, 5, 19, 24, 43, 59, 66, 104, 105, 120
Reverberating the wild, 40–41
Re-wilding, see Wild pedagogies
Rigour, x, 42
Rocks, seeing, 113
Rose, Deborah Bird, 13

S

Sami people, 33, 46n19
Saul, John Ralston, 18
Sault College, Canada, 124
School, vii, viii, xi, 4, 18, 38, 43, 69, 82, 86, 91, 96, 110, 114, 117, 121, 129
   starting age, vii
Science, 52
Science and technology, viii
   experimentation, 60
Scotland, xii, 7–17, 29, 30, 41, 45, 60, 71, 72, 79, 82, 98, 117, 120, 124, 127, 128
Seals, 9, 38, 57, 72, 73, 103
Self, examining the, 93
Self-regulation, 66
Sensory awareness, 55
Serres, Michel, 53, 54, 61n6
"Sharing Spaces" policy, 125
Sheridan, Joe, ix, x, xxix
Silurian epoch, 97
Simon Fraser University, Canada, 114, 127
Snyder, Gary, 44
Socio-cultural change, see Cultural change
Soil loss, 3
Species loss, see Extinctions
Spontaneity, 84–88, 114
Staffa, Isle of, 17, 41, 42, 84, 85
Sterling, Stephen, 6
Surprise, 40, 68, 84, 112, 114
Sustainability education, 66, 67

T

Teachers
   habits, 101
   rebel, 64
Testing, educational, x, 65, 68
Thunder Bay Art Gallery, Canada, 111

Time
  clock-, 92–94
  deep, 54, 78, 91, 94
Touchstones, xii, 18, 20, 24, 74, 78–106, 110, 118, 125, 127, 129
  questions, 121, 125, 129
Trees for Life, 117–119, 121
Treshnish Isles, 79
"Trouble with Wilderness, or Getting Back to the Wrong Nature, The", 25
Tuskegee Institute, 116

U

UNESCO, 5, 6
United Nations Organisation, 6
Unknown, 69, 84–88, 98
  See also Not-knowing
Urban spaces, 88, 91

V

Victoria, Queen, 13, 30
Voices, x, xii, 7, 11, 12, 16, 18, 34, 37–39, 72, 74, 79, 81–83, 87, 89, 90, 96, 101–104, 113, 115, 119, 127
Vuntut Gwitchin First Nation, 31

W

Walking, ix, 68, 72, 80, 110, 111, 120, 121
Wals, Arjen, 65
Washington, Booker T., 116
Welz, Andrea, 124
"Wet desert", 13

Wild
  learnings, 128
  locating, 88–91
  research, 127
  term, 46n19
Wilderness
  agency, 37, 40
  characterization of, 25, 27, 31, 40
  civilization and, 33–35
  concept of, 25, 26, 31, 33, 35
  critiques of, 13, 15, 25, 34
  definition of, 24, 25
  education and, 19
  etymological roots, 29
  pastoral view of, 13
  protecting, 29, 30
  re-thinking, 2, 25, 29, 40
  true, 25
  voice, 34
Wildness, x, 2, 13, 20, 24, 27, 30, 32, 43, 44, 65, 66, 69–70, 89, 102, 115, 119, 122, 128
"Wildoerness", 26, 32
Wild pedagogies, 69
  aims, x, 12–13
  art and (*see* Arts)
  conferences and gatherings, xi, 8, 26
  graduate course, xi, 111, 124
  history, 24, 71
  pluralization of, xi
  practice, 78–106, 116, 125
  re-wilding of education, x, 104, 117–122
  strategies for, 2
  term, viii
  touchstones, 24, 78–106, 110, 121, 125, 127
  Victorian concept of, 32

Wild Pedagogies: A Floating Colloquium (Yukon, 2014), xi, 111, 114, 117, 122, 124
Wild Pedagogies: A Sailing Colloquium (Scotland, 2017), xii, 7–17, 117, 121, 124
Wild Pedagogies: The Tetrahedron Dialogues (2016), xi
Wolves, 111, 118–121
Wonder, 3, 20, 27, 31, 34, 35, 40–44, 68, 85, 90, 94, 105, 111–113, 124, 128

Wood-Alexander, Vivian, 111
World Environmental Education Congresses, xi
Writing, collaborative, 12

Yellowstone National Park, USA, 30
Yukon First Nations, 31
Yukon River, Canada, xi, 7, 114–117
Yukon Territory, 31

Lightning Source UK Ltd.
Milton Keynes UK
UKHW022345181118
332479UK00012B/94/P